Laccoliths;
Mechanics of emplacement and growth

Charles E. Corry
Department of Geophysics
Texas A&M University
College Station, Texas 77843

SPECIAL PAPER

220

Published by The Geological Society of America, Inc.
3300 Penrose Place, P.O. Box 9140, Boulder, Colorado 80301

GSA Books Science Editor Campbell Craddock

Printed in U.S.A.

Library of Congress Cataloging-in-Publication Data

Corry, Charles E., 1938–
    Laccoliths : mechanics of emplacement and growth / Charles E.
Corry.
    p.   cm. — (Special paper / Geological Society of America ;
220)
    Bibliography: p.
    ISBN 0-8137-2220-9
    1. Laccoliths.   I. Title.   II. Series: Special papers (Geological
Society of America) ; 220.
QE611.C67   1988
552'.3—dc19   ∧ 88-10166
                         CIP

Cover photo: Table Mountain in the Henry Mountains, Utah, is
typical of punched laccoliths. The near vertical flanks and flat top
are the result of an intrusion which mechanically punched its way
through the roof rock. The view is of the northwest side of the
laccolith. Photo by Charles E. Corry.

Contents

Acknowledgments

In attacking a problem of such broad scope, I have called heavily on the individual expertise of many people, and their guidance has been invaluable. Davis A. Fahlquist, Department of Geophysics, provided invaluable assistance in obtaining financial aid and in the analysis of the gravity data presented. John Handin, John Logan, and George M. Sowers (deceased), of the Center for Tectonophysics, served as invaluable references for the problems associated with the mechanical behavior of rock. Robert Scott, now with the U.S. Geological Survey of Denver, Colorado, gave a great deal of his time to discussions of igneous petrology and intrusion mechanics. James A. Stricklin (deceased), and his colleague Walter E. Haisler, in the Aerospace Engineering Department, provided the background and assistance required to enable me to do the included finite element analysis.

I have had the willing aid of several field assistants, and I would like to express my gratitude for their help. Roberta Hoy and Ruth A. Corry acted as field assistants during the survey of the Solitario area in Texas. Kenneth Cooper served as my field assistant for the surveys in the Henry, Abajo, and La Sal Mountains of Utah. Shigeo Iwamura aided me during the surveys in the Sundance area of Wyoming and the Little Belt Mountains, Montana. Greg Pekar gave up his Christmas vacation to help with the survey of Big Bend National Park, Texas.

Access to the Solitario was by the kind permission of Phelps Anderson and Ralph Hager of the Big Bend Ranch, Diamond A Cattle Corporation. Most of the other areas surveyed are on federal lands, and the cooperation of the many officials of the U.S. Forest Service, Bureau of Land Management, and National Park Service is gratefully acknowledged. Frank Deckert, with the National Park Service, Big Bend National Park, Texas, was particularly helpful during the survey of that area.

The gravity survey of the Solitario was partially financed by NASA grant NGL 32-004-011 through the University of New Mexico, and a grant from the Bureau of Economic Geology, Austin, Texas. The surveys of the Henry, Abajo, and La Sal Mountains were partially financed by a Grant-in-Aid of Research from Sigma Xi, the Scientific Research Society of North America, and a matching grant from the Research Committee of Texas A&M University. The survey of Big Bend National Park was partially financed by a grant from the Bureau of Economic Geology, Austin, Texas. The finite element analysis was partially financed by a grant from the American Association of Petroleum Geologists and through NASA contract NAS 8-31332.

I would like to thank J. D. Hendricks, with the U.S. Geological Survey, for permission to review his unpublished gravity data from the San Francisco Mountains, Arizona, and the help of E. W. Wolfe, of the U.S. Geological Survey in Flagstaff, Arizona, for his aid in locating a copy of Hendrick's unpublished map.

Portions of the manuscript were read by Anthony Gangi of Texas A&M University, Paul Proctor of Brigham Young University, and S. Kerry Grant of the University of Missouri at Rolla. While I do not agree with his models, the comments of Charles B. Hunt and his co-worker, Irving J. Witkind, in their reviews of this paper, were very helpful in clarifying my own models. The review by Donald W. Hyndman was invaluable, and I would like to express my thanks for his making available to me a preprint of his and Dave Alt's work in the Adel Mountains, Montana. David D. Pollard very kindly sent me a preprint of his work with Marie Jackson in the southern Henry Mountains, and discussions with Pollard and his students over the years have strongly influenced many of my ideas regarding laccoliths. The help and comments of Campbell Craddock, Books Science Editor for the Geological Society of America, were an essential part of completing this manuscript. A great deal of the work that went into compiling the included gazetteer was done by Ruth A. Corry, for which she has my heartfelt thanks. However, errors of translation, omission, and commission are solely my responsibility. The conclusions are my own.

Dedication

Dedicated to George M. Sowers

1921–1974

A great many of the ideas presented here originated in delightfully raucous debates with Mr. Sowers. The inspiration and guidance he provided is gratefully acknowledged and sorely missed.

George M. Sowers was born on December 17, 1921, in Ridgefield, Connecticut. He completed a B.A. degree in chemistry at Wesleyan University in 1942, and an M.A. degree in structural geology at Johns Hopkins University in 1944, after which he went to work for the U.S. Geological Survey as a field geologist. It was during this period that he developed his interest in laccolithic intrusions while working in the Little Belt Mountains in Montana. He continued that interest until his death. After leaving the Survey in 1946, he returned to graduate school at Columbia University where he completed all requirements for the Ph.D. except the dissertation. He became an assistant professor at Washington University in St. Louis in 1948. In 1951 he went to work for Shell Development Company in Houston, Texas, as a research geologist. At Shell he joined John Handin's rock mechanics group. When Handin's work at Shell was phased out in 1966, Sowers went with the group to Texas A&M University, where he helped found the Center for Tectonophysics in 1967. As a lecturer and research associate, he worked as the principal theoretician for the Center, an accomplishment all the more impressive in that his knowledge of mathematics was largely self-taught.

He was killed in an automobile accident near Kingsville, Texas, in December 1974, while leading a field trip to Central America with his graduate students.

ABSTRACT

Gilbert (1877) proposed that the level of emplacement of laccoliths is controlled by the density contrast between rising magma and the weighted mean density of the overburden. For felsic laccoliths, his hypothesis is strongly supported by gravity surveys of a number of laccolith groups. Epizonal felsic laccoliths are consistently found to have zero density contrast with the host rocks. Constraining the emplacement level provides a basis for analysis of the growth of laccoliths.

Mechanical analysis suggests that the diverse shapes of laccolithic intrusions observed in the field can be represented by a continuous series of intrusion modes between two distinct end members. The simplest end member is an epizonal intrusion formed by a single sill that acts mechanically as a vertical punch. Punched laccoliths are characterized by flat tops, peripheral faults, and steep or vertical sides.

The other end member results from the intrusion of multiple sills stacked vertically in a fashion suggestive of a Christmas tree. The multiple-level loading results in plastic deformation of the country rock. Christmas-tree laccoliths lack peripheral faults and have a characteristic rounded dome appearance on the surface. The floor of these laccoliths may, or may not, sag. Gilbert's (1877) ideal laccolith falls between these two end members.

The end members of the laccolith growth series are treated as boundary value problems in continuum mechanics. Geometrically and materially nonlinear finite element analysis is used to solve the boundary value problems. Field observation, a physical model, and the theoretical models provide convergent answers to the mechanical analysis of the growth of laccoliths.

As a check on the theoretical models, a gazetteer of the dimensions and locations of approximately 900 laccoliths is included. Of these, approximately 600 are located in the United States. If North America represents a statistically valid sample, then there must be between 5,000 and 10,000 laccoliths around the world.

CHAPTER 1

INTRODUCTION

The conceptual simplicity of laccoliths has invited attempts at quantitative analysis since their genesis was first described in a classic report on the geology of the Henry Mountains of Utah by G. K. Gilbert in 1877.

In my study of laccoliths, I have had the advantage of modern maps, computers, and transportation (though horses still played a basic role in the surveys), and my investigations of laccoliths have been, perhaps, more wide ranging. To test Gilbert's thesis I have used an interaction between geophysical methods of investigating geologic structures, field geology, and computer and physical models of laccoliths. I do not differ in any important way with Gilbert's original conclusions. This is remarkable only in that nearly all other investigators in the intervening century have differed substantially with Gilbert's interpretation.

DEFINITION OF LACCOLITHS

Gilbert (1877, p. 19) defined laccoliths (Gr., stone cistern) on the basis of their mode of formation. In his usage, laccoliths are formed by magma that "insinuated itself between two strata, and opened for itself a chamber by lifting all the superior beds." As reviewed by Daly (1933, p. 81), there is general consensus on the following characteristics of laccoliths. (a) Laccoliths are formed by forcible intrusion of magma and initially are entirely enclosed by the invaded formations except along the relatively narrow feeding channel. (b) Like sills, laccolith contacts commonly follow a bedding plane, though many instances are known where the intrusion cuts across bedding. The plane of the intrusion remains sensibly parallel to the Earth's surface at the time of the intrusion, although the floor of large intrusions may sag as the intrusion grows. (c) In cross section, the ideal laccolith of Gilbert (Fig. 1) has the shape of a plano-convex, or doubly convex, lens flattened in the plane of the bedding of the invaded formation. The lens may be symmetrical or asymmetrical in profile; circular, elliptical, or irregular in plan view. (d) There is a complete gradation between sills and laccoliths, with no clearly defined point at

which a sill becomes a laccolith. (e) A laccolith lifts its roof as a result of the forcible injection of magma. Hunt and Mabey (1966) point out that the presence of concordant roof pendants, which are borne upward from shallow depths, implies a floored, laccolithic intrusion.

Intrusions that have apparently gone through the forcible intrusion and thickening process are referred to in this paper as laccolithic intrusions, or more simply as laccoliths, regardless of final cross-sectional form.

Billings (1972) uses the arbitrary distinction that an intrusive body is a laccolith if the ratio of the diameter to the thickness is less than ten, and the body is a sill if the ratio is greater than ten. There has been no consistent usage of his definition in the literature. As will be shown, I can find no basis, either mechanically or from field observation, for using Billings' distinction between a sill and a laccolith.

While my principal objective is understanding laccolithic intrusions by analyzing reasonable mechanical models, as begun by Gilbert (1877), it is desirable first to review the descriptive terms that have evolved as a consequence of the wide variety of forms of laccolithic intrusions found.

Lopoliths: A class of laccoliths

In his definition of lopoliths (Fig. 2), for which the Duluth gabbro is the type example, Grout (1918, p. 516) points out that the gabbro ". . . has as definite a roof and floor as a laccolith or sill, and was intruded along a surface approximately corresponding to a previous structure,—the unconformity at the base of the Keeweenawan." After reviewing and eliminating the possibility that the Duluth gabbro may be an ethmolith or chonolith, he concludes (p. 516–517) ". . . that by a process of elimination the gabbro is placed with the laccoliths." Grout then goes on to point out that, contrary to the accepted form of a laccolith, with a raised roof and flat floor, that the Duluth gabbro forms a basin as shown in Figure 2, and the floor and roof dip inward. He then named this type of forcible intrusion a lopolith. Modern workers (e.g.,

C. E. Corry

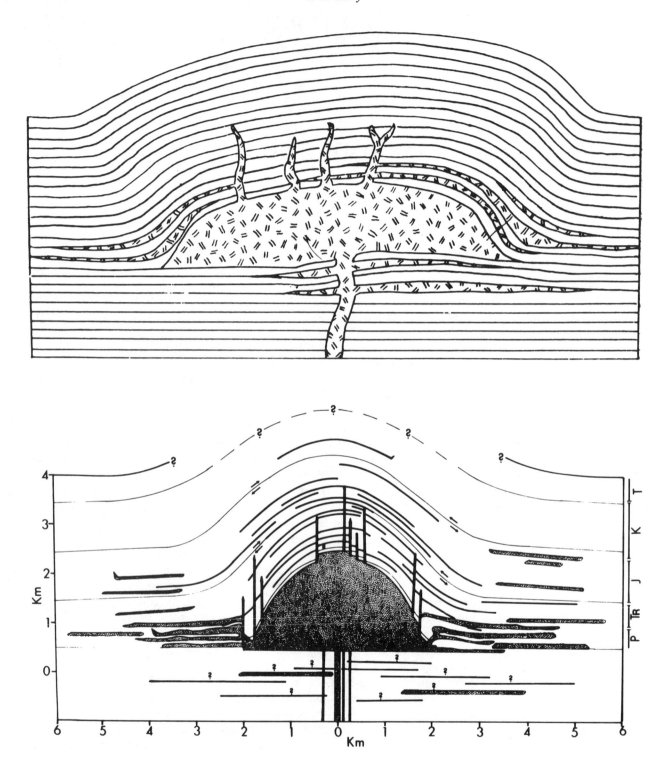

Figure 1. The ideal form of a laccolith (top) as envisioned by Gilbert (1877). As shown on the bottom, Jackson and Pollard (1988) have revised Gilbert's ideal form after more detailed mapping in the Henry Mountains, Utah.

Wager and Brown, 1967, p. 542; and Bridgwater and others, 1974, whose idealization is shown in Fig. 3) agree that initially the roof was elevated by the intrusion and the central part may be depressed by post-intrusion sagging. Sagging in crustal rocks implies plastic deformation, which requires elevated pressure and temperature; hence, deep levels of intrusion are required. Mesozone pressures and temperatures are in accord with the field observations of Bridgwater and others (1974, p. 66). Lopolith-like intrusion has been extensively modeled by Ramberg (1981). It thus appears that magmas that are forcibly injected in the epizone will tend to form laccoliths whose floors remain relatively flat, while magmas injected in the mesozone—or possibly the catazone—will tend to form lopoliths in which the floor sags after, or possibly during, injection of the magma.

There is no clear-cut distinction between a laccolith and a lopolith. Lopoliths were uniformly called laccoliths prior to Grout's (1918) paper. I include both in the definition of laccolithic intrusions and refer to both of them as laccoliths in a generic sense unless there is a reason to make a distinction.

Chonoliths

The term chonolith (Fig. 4) was introduced by Daly (1914, 1933) to describe forcible intrusions with complex structural relationships that do not readily fit the category of a laccolithic intrusion. Such irregular intrusions do not conveniently fit any classification scheme, and the term is still useful. The use of chonolith to describe an intrusion has the value of indicating that the investigator examined the structural relations. Chonolith would seem to be preferable to the word "pluton" when describing structural relationships of forcible intrusions that are complex or not presently understood.

Additional names for laccolithic intrusions

Other descriptive terms for laccolithic intrusions, and variants thereof, include: akmolith, bysmalith (Fig. 5), cactolith, ductolith, ethmolith, harpolith, phacolith (Fig. 6), sphenolith, trap door, and Christmas-, compound-, or cedar-tree (Fig. 7) laccolith. The definitions of these terms are given in Appendix A.

Because of its utility and imagery in describing the mechanics of one end member of my models, I have used the Christmas-tree laccolith name. Otherwise, unless a descriptive term is required for clarity, or to illustrate a point, I have used mechanical terms (e.g., punched laccolith) or laccolith in the generic sense for any or all of the intrusions described above.

Laccolith group

A useful term in describing laccolithic intrusions is a "laccolith group." The term was first used by Gilbert (1877) and was more formally used by Cross (1894) in describing clusters of laccoliths on the Colorado Plateau. The term implies a spatial relationship between a number of laccoliths. Laccoliths within a

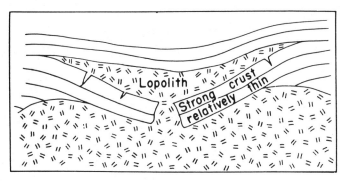

Figure 2. Lopolith, as envisioned by Grout (1918).

group may also be related by time of intrusion and petrology of the igneous rocks in the group.

HISTORY OF THE LACCOLITHIC CONCEPT

Gilbert's work

The section of interest is Chapter IV (Gilbert, 1877, p. 51–92) on "The Laccolite." Gilbert started by offering proof that the igneous rocks of the Henry Mountains are intrusive rather than extrusive. The concept was revolutionary when introduced, but today seems so well proven that no additional comment is required.

Gilbert then discussed the facts his hypothesis must account for and gives a general description of laccoliths. His comment on the layered structure of several of the laccoliths, reinforced with sketches of layered laccoliths, formed a mechanical basis for my concept of Christmas-tree laccoliths.

Neutrally buoyant elevation. Gilbert noted that the intrusive rocks of the Henry Mountains group were all of one type, and this has been substantiated by Hunt and others (1953). Gilbert then approached the fundamental problem of the cause of laccolithic intrusions. The present research supports the following hypothesis (Gilbert, 1877, p. 66–69) for laccolith emplacement at the neutrally buoyant elevation:

It is not necessary to broach the more difficult problem of the source of volcanic energy. We may assume that molten rock is being forced upward through the upper portion of the earth's crust, and disregarding its source and its propelling force may restrict our inquiry to the circumstances which determine its stopping place.

Let us further assume, but for a moment only, that the cohesion of the solid rocks of the crust does not impede the upward progress of the fluid rock, nor prevent it from spreading laterally at any level. The lava will then obey strictly the general law of hydrostatics, and assume the station which will give the lowest possible position to the center of gravity of the strata and lava combined.

(1) If the fluid rock is less dense than the solid, it will pass through it to the surface and build a subaerial mountain.

(2) If the upper portion of the solid rock is less dense than the fluid,

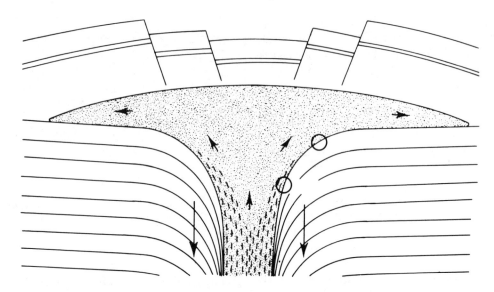

Figure 3. Idealized cross section of the rapakivi granite intrusions of South Greenland from Bridgwater and others (1974). The country rock is uplifted above the intrusion, and a crestal graben is formed. Beneath the intrusion the country rock is downwarped and individual units are attenuated close to the margin of the intrusion. The arrows show the direction of movement for both the magma and the country rock. Bridgwater and others' (1974) results closely follow the theoretical predictions of Ramberg (1981). The idealization is based on field observations by Bridgwater and others (1974) of several intrusions in South Greenland.

while the lower portion is more dense, the fluid will not rise to the surface, but will pass between the heavy and light solids and lift or float the latter.

(3) If the crust be composed of many horizontal beds of diverse and alternating density, the fluid will select for its resting place [*the neutrally buoyant elevation, comment added*] a level so conditioned that no superior group of successive beds, including the bed immediately above it, shall have a greater mean specific gravity than its (the fluid's) own; and that no inferior group of successive beds, including the bed immediately beneath, shall have a less mean specific gravity than its own.

. . . the first case is that of a volcano; the second is that of a laccolite; and the third is the general case, including the others and applying to all volcanoes and laccolites.

. . . In brief, since lavas are fluid they are subject to the law of fluid equilibrium, and their behavior is conditioned by the relations of their densities to the densities of the solids which they penetrate; and since the latter solids are rigid and coherent, it is further conditioned by the resistance which is opposed to their penetration. When the resistance to penetration is the same in all directions, the relation of densities determines the stoping place of the rising lava; but when the vertical and lateral resistances are unequal, *their* relation may be the determining condition.

. . . We are then led to conclude that the conditions which determined the results of igneous activity were the relative densities of the intruding lavas and of the invaded strata; and that the fulfillment of the general law of hydrostatics was not materially modified by the rigidity and cohesion of the strata.

Gilbert then proceeds to develop a mechanical model for laccoliths, which is preserved in my punched model. Gilbert also noted that there are no small laccoliths (<1 km in diameter) and

provided quantitative reasons. That observation has been well substantiated by all subsequent field work.

Depth of intrusion versus laccolith diameter. One of the few observations Gilbert makes that I have been unable to verify with field evidence is a relation between depth of intrusion and diameter of the laccoliths. The concept is intuitively appealing. In the included theoretical models of punched laccoliths, a definite relation was found between depth of intrusion and diameter of the laccolith. However, for Christmas-tree laccoliths, a depth-diameter relationship is obscure, and it will be particularly obscure to the field observer who usually has little information about which level of the intrusion has been exposed by erosion.

Gilbert (1877, p. 88) makes an astute point. ". . . It is always hazardous to attempt the quantitative discussion of geological problems, for the reason that the conditions are apt to be both complex and imperfectly known. . . ." I proceed then only with a degree of temerity.

Other early workers

Gilbert's work initiated a flurry of research into intrusive mechanisms. The greatest controversy centered around the effect of magma viscosity. The debate over the role of magma viscosity began with Weed and Pirsson (1898). Paige (1913) and Darton and Paige (1925) purported to explain the formation of bysmaliths (Iddings, 1898) by a progressive increase in the viscosity of the magma. Pollard and Johnson (1973) have shown, using me-

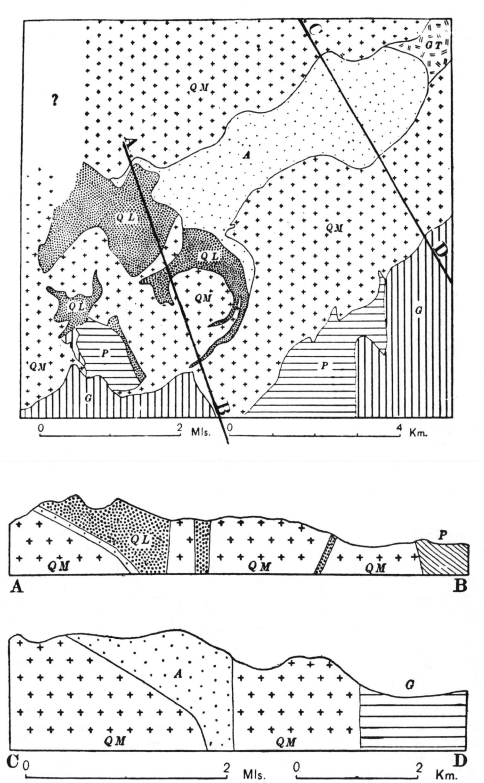

Figure 4. Plan and cross sections of chonoliths in Monarch and Tomichi districts, Colorado. After Crawford (1913) and Daly (1933). At top is plan view of chonoliths of quartz latite porphyry (QL) and andesite (A) cutting quartz monzonite and quartz monzonite porphyry (QM), post-Carboniferous granite (GT), Paleozoic sediments (P), and Precambrian granite (G). Below are sections along the lines A-B and C-D above. Underground contacts partly determined by mining.

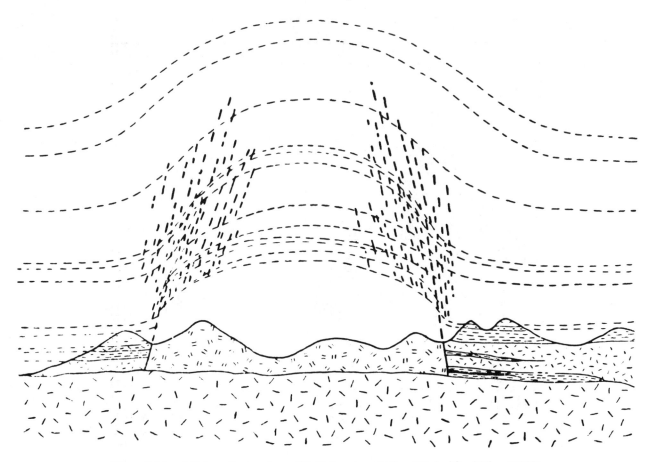

Figure 5. Mount Holmes, the type bysmalith, Yellowstone National Park. After Iddings (1898).

chanical arguments, that Paige's model is inaccurate. Early authors required a high magma viscosity to account for the observation that large inclusions in the magma do not sink. Measurements in basaltic lava lakes (Shaw and others, 1968; McBirney and Murase, 1984) since then have demonstrated that a magma containing crystals may behave as a Bingham substance and, hence, the magma has a finite shear strength. Pollard and Johnson (1973) estimated that a shear strength of $\geqslant 100$ Pa is sufficient to support the largest observed inclusions in the laccoliths of the Henry Mountains, Utah.

Most of the papers immediately following Gilbert's initial study serve to point out the diversity of shapes laccolithic intrusions assume and introduced a plethora of descriptive names (see Appendix A). Hunt and others (1953, p. 139–151) have reviewed these early studies. It is of interest to note that some laccoliths breached their roof, reached the surface, and there spread laterally, usually to a very limited extent (Stark, 1907, p. 52–56, and 1912, p. 10–80; Lachmann, 1909, p. 336; Cornu, 1906, p. 45–46; Robinson, 1913, p. 74–85; Mintz, 1942; Lovejoy, 1976, p. 59).

In 1925, MacCarthy published a review of the facts and theories concerning laccoliths known at that time. As MacCarthy

(1925, p. 9) points out ". . . in all cases noted the contact-metamorphic effect of laccoliths are much less than is usual with other intrusive bodies of equivalent magnitude." It should be noted, however, that mafic laccoliths have much more distinct thermal metamorphic halos than do felsic laccoliths, due to the higher temperatures of basic magmas. MacCarthy (1925) also notes that laccoliths are usually (he says invariably, but that is not true) intruded into shale horizons or along planes of unconformity. MacCarthy (1925, p. 9) also undertook an extensive series of qualitative experiments on the growth of laccoliths, which were suggestive of many field relations I had seen. His models stimulated several debates with Mr. G. M. Sowers on the mechanics of laccolithic intrusions, which in turn, led to the development of the models presented here.

Modern work

Much of the modern research on laccoliths was begun by the U.S. Geological Survey in the period following World War II. Laccoliths appeared to have a genetic relationship with uranium ores, and the U.S. Geological Survey was at the time engaged in a large-scale search for this strategic metal. Their field studies are tabulated in the included gazetteer (Appendix B).

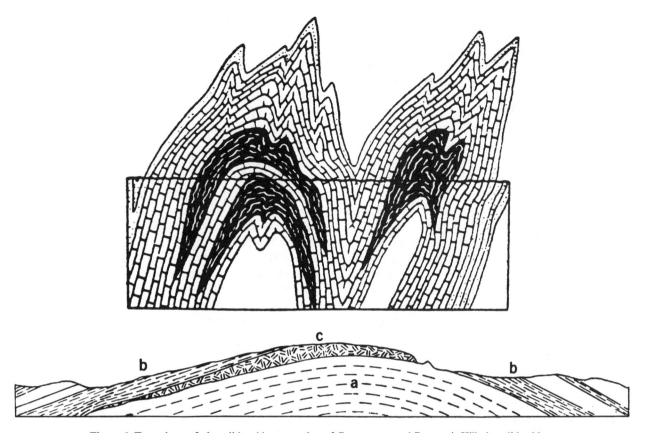

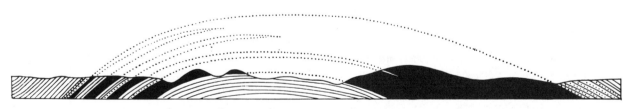

Figure 6. Two views of phacoliths. Above, section of Gouverneur and Reservoir Hill phacoliths, New York, with reconstruction above the existing erosion surface. White lines on the black of the granitic phacoliths represent foliation. After Buddington (1929). Below, cross section of a phacolith in anticline of Ordovician strata, Corndon, Shropshire, England. a, Mytton flags and Hope Shale; b, Stalely ashes and andesite; c, dolerite. After Lapworth and Watts (1894).

Figure 7. Christmas-tree laccolith by the Moldau River (Vitava), in central Czechoslovakia. The section is about 5 km long. The vertical scale is exaggerated 25 percent. After Kettner (1914).

The use of continuum mechanics for the study of laccoliths was begun by Pollard (1969). Johnson (1970), Savage and Sowers (1972), Johnson and Pollard (1973), Pollard and Johnson (1973), Pollard (1973), Savage (1974), Johnson and Ellen (1974), Pollard and Muller (1976), Corry (1976), Koch and others (1981), and Jackson and Pollard (1988) have continued the use of continuum mechanics.

Geophysical investigations of laccoliths, principally gravity and magnetic surveys, were begun by Greenwood and Lynch (1959). Case and others (1963) surveyed the La Sal Mountains, Utah, and later extended their survey over the Four Corners region (Case and Joesting, 1972). Cook and Hardman (1967) presented a regional gravity survey of the Iron Springs and Pine Valley area of southern Utah. Peterson and Rambo (1967, 1972) made a gravity survey of the Bearpaw Mountains, Montana. Kleinkopf and others (1972) made a gravity survey of the Little Belt Mountains, Montana; a survey that I supplemented. Kleinkopf and Redden (1975) made a gravity and aeromagnetic survey of the Black Hills in South Dakota and Wyoming. Again, I have supplemented their survey. Plouff and Pakiser (1972) made a gravity survey of the San Juan Mountains, Colorado. Bothner (1974) presents a gravity study of the Exeter pluton in New Hampshire. Hendricks (personal communication, 1975) has done a gravity study of the San Francisco Mountains, Arizona. Hunt and Mabey (1966) include a gravity and magnetic survey in their report on the geology of the Death Valley area, California. Briden and others (1982) made a gravity and magnetic map of the Channel Islands, Great Britain, an area that contains a number of laccoliths. These gravity surveys are discussed in detail in a subsequent chapter.

Large laccolithic intrusions composed of mafic and ultramafic rock have significant positive density contrasts. Gravity and magnetic surveys have been used to map a number of such features. The Trompsberg lopolith in South Africa was defined by gravity and magnetic surveys (Buchmann, 1960). Laccoliths are defined by gravity surveys near Mara in northern Tanzania (Darracott, 1974). Rusu (1975) made a gravity survey over laccoliths in the central and southern areas of the Harghita Mountains of Romania. However, no gravity map is included in his paper. Bermingham and others (1983) used a gravity survey to define the Darfur dome in western Sudan. An electrical resistivity survey (Davino, 1980) was used to map diabase laccoliths in the state of São Paulo, Brazil. Nodop (1971) used seismic refraction to study the structure of laccolithic intrusions in Germany.

In addition to the supplemental surveys in Montana and Wyoming mentioned above, I have conducted gravity surveys of laccolithic areas in Colorado, Texas, and Utah. These surveys are discussed below.

TABULATION OF KNOWN LACCOLITHS:
A GAZETTEER

Before attempting a quantitative analysis of laccoliths, the available data on laccolithic intrusions were reviewed. The results of my search are tabulated and presented as a gazetteer in Appendix B. The references given in the gazetteer are generally limited to authors who have discussed the structure of the intrusion. Papers limited to discussions of sedimentary petrology, stratigraphy, paleontology, or similar topics, are particularly likely to have been missed, especially if not cross-referenced under laccoliths in the American Geological Institute GEOREF database.

Because of the plethora of names (above and Appendix A) used in the geological literature for laccolithic intrusions, I have generally not attempted to distinguish the type of laccolith in the gazetteer. I have not been critical of what may, or may not, be termed a laccolithic intrusion. If any author, at any point, said the intrusion may be a laccolithic intrusion, I have included it in the gazetteer. Some bodies presently included in the gazetteer will prove not to be laccolithic intrusions at all. For example, Sudbury, in Ontario, Canada, was originally called a laccolith, then a lopolith, and now the evidence suggests that it is an impact structure. However, the number of structures included in the gazetteer

that prove not to be laccoliths will be small compared to the number of laccoliths yet to be described in the geological literature.

Unless large-scale topographic maps were available to me (a rarity), I have depended for locations on the gazetteers of the U.S. Board on Geographic Names prepared by the Defense Mapping Agency (DMA) Topographic Center. Since many geographic names are not unique, and frequently the geographic information contained in the articles was insufficient to pinpoint the location of the laccolith within 10° in latitude or longitude, many locations may be in error. I have had difficulty locating laccoliths described in the foreign literature. Thus, though a number of laccoliths have been described in South America by Steinmann (1910) and Erdmannsdörfer (1924), I have not included them in the gazetteer because I was unable to locate the laccoliths from their descriptions, and the names of the features they used were not given in any gazetteer available to me. Various factors, such as language, access to foreign publications and maps, and others, have limited the number of laccoliths tabulated for foreign countries. The gazetteer is as complete and up to date (mid-1987) as feasible, but it certainly contains errors of omission, translation, and transcription.

Approximately 900 laccoliths, of which about 600 are located in the United States, have been tabulated. Only 38 countries are included in the tabulation, and these are primarily economically developed countries, which have been geologically mapped to some extent. The gazetteer thus seriously understates the number of laccoliths that must exist in the world. However, it is possible to extrapolate from the available data. Assuming that not more than two-thirds of the laccoliths in the United States have been described in the geologic literature, and a lesser percentage in Mexico and Canada, I suggest that the actual number of laccoliths in North America exceeds 1,000. The world total must then be between 5,000 and 10,000, unless laccoliths are inexplicably abundant only in North America.

Size information has usually been included only if explicitly stated by the author(s). The thickness of the laccoliths given in Appendix B may have been measured directly if the floor and roof are visible or determined from the measured structural relief. I have not distinguished which method was used, since either technique is subject to about the same degree of error.

The most extensive compilation of radiometric ages for laccolithic intrusions I have found is in Henry and others (1986), which covers Trans-Pecos Texas. Corry and others (1988) made a detailed study of the age relations of the intrusive sequence within a single laccolith, the Solitario, in the Big Bend group, Texas. Armstrong (1969), Cunningham and others (1977), and Sullivan (1987) have provided radiometric ages for laccoliths of the Colorado Plateau. Elsewhere, data are usually insufficient to clearly establish time or petrologic relationships. However, most of the laccoliths tabulated in the gazetteer are Tertiary in age and felsic in composition. Whether the laccolithic type of intrusion became more common in the Tertiary, or whether older laccoliths have simply been eroded away, is unknown. The data suggest that those laccoliths older than Tertiary are more mafic than more recent laccoliths. McBirney (1984, p. 188) also notes that most of the largest mafic intrusions are of Precambrian age and that younger intrusions are generally much smaller.

Group names given in the gazetteer (Appendix B) are of my own invention. References for a group are given in chronologic order of the investigations done in that group.

In reviewing the literature on laccoliths, it appeared to me that two possible misconceptions exist with regard to laccoliths. The first apparent misconception, and of greatest economic importance, is the relation of laccoliths to ore deposits. The relation between metallic ores and laccoliths was first recognized by Storms in 1899. Many modern geologists, at least in print, regard laccoliths as unlikely sites for ore deposits. In a monograph on the relations of ore deposition to doming, Wisser (1960) does not once use the term laccolith. He refers to such domes as "boils," "blisters," and "tumors." The igneous rock of the laccolith itself is almost invariably barren of ore in commercial concentrations. However, fractures in the laccolith or country rock, and the breccias resulting from the formation of the laccolith, have very frequently acted as sites for economic concentrations of ore minerals from hydrothermal activity postdating the laccolith (e.g., see Hodgson and others, 1976). Thus, the mining geologist will recognize a large number of well-known mining districts in the gazetteer, among them the Black Hills of South Dakota and the Terlingua district in Trans-Pecos Texas. Economic concentrations of substances other than ore minerals have also been found associated with laccolithic domes. Most recent of these is the discovery of large quantities of carbon dioxide (CO_2) in the Spanish Peaks and Sleeping Ute Mountains of Colorado. The CO_2 is being used in tertiary oil recovery.

A second possible misconception is the idea that an intrusion with discordant or faulted flanks must be a stock. Gilbert (1877) appears to have recognized that the margins of shallow intrusions will be faulted in some circumstances. Laccoliths with faulted or discordant margins were also the subject of considerable debate in the earlier literature (MacCarthy, 1925). Hunt and Mabey (1966) and Jackson and Pollard (1988) discuss the problems of recognizing a stock versus a laccolith with faulted margins. Iddings (1898) coined the term bysmaliths for this type of laccolith. I use the term "punched" laccolith to describe the mechanics of the process and avoid the debate engendered by Paige (1913) and Darton and Paige (1925) regarding the role of magma viscosity in the formation of these types of laccolithic intrusions. Johnson and Pollard (1973) have shown from mechanical arguments that their model is invalid.

STATEMENT OF PROBLEMS

The problems associated with the mechanics of emplacement and growth of laccoliths can be stated explicitly. In tabular form they are:

1. What is the driving force (or forces) for the intruding magma?

2. What factors determine the level of intrusion and cause the magma to reorient from dominantly vertical motion to radial spreading as a horizontal sheet?

3. What physical parameters affect the process of intrusion? If these parameters vary during the intrusion process, which are dominant at any given time?

4. What limits the radial horizontal spreading and what initiates large-scale thickening of the magma to form the characteristic laccolith shape?

Quantitative models are developed in this paper for these problems. The principal objective of this work is understanding the mechanics of laccolithic intrusions. The notation required to define the variables associated with the intrusion mechanics is given in Appendix C.

In addition to their conceptual simplicity, laccoliths lend themselves to an attempt at quantitative interpretation because of the following characteristics:

a. Laccoliths are common, and various stages of erosion within a group allow direct field observation of the relationship between the intrusions and the country rock.

b. Their size, usually in a range from 1 to 10 km in diameter, makes rapid investigation possible.

c. Laccoliths exhibit relatively simple geometries, with radial or ellipsoidal symmetry simplifying theoretical analysis.

d. The deformation accompanying the growth is dominantly mechanical for epizonal laccoliths, as evidenced by very limited zones of contact metamorphism. Hence, thermal and chemical modification of the properties of the country rock by the invading magma can be safely ignored.

e. Previous investigators of the field relationships, theoretical studies of the mechanics of intrusion, and laboratory studies of physical models of laccoliths have provided a strong foundation for development of quantitative models to describe the emplacement and growth of laccoliths.

An important parameter in the development of the models is my own experience. I have looked at perhaps 200 laccoliths in Colorado, Montana, New Mexico, Texas, Utah, and Wyoming. Nearly all of the laccoliths I have seen can be characterized as felsic intrusions. Of the approximately 200 laccoliths visited, I have examined perhaps three dozen in some detail, usually at exposures pointed out by previous workers in the area.

In formatting the following models, I may have oversimplified the problems and ignored critical parameters, for example, volatile content. My only defense is that those parameters appear to have such a negligible effect on the mechanical system, which I am investigating, that I am unable at present to isolate them from the gross aspects of the intrusion.

PUNCHED AND CHRISTMAS-TREE LACCOLITHS

Gilbert (1877) described a single type of laccolith, although he did recognize that laccoliths may be stacked one above another. I suggest that the various forms of laccolithic intrusions observed in the field are, for mechanical reasons, part of a contin-

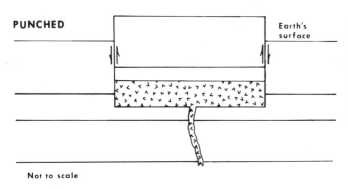

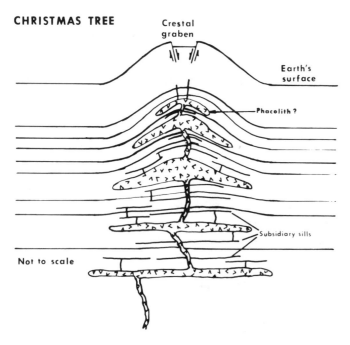

Figure 8. Punched and Christmas-tree laccoliths.

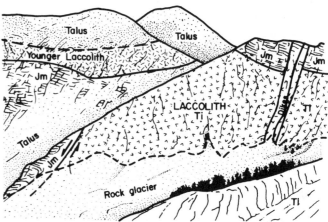

Figure 9. The Mount Peale–Mount Tukuhnikivatz laccolith at the head of Brumley Creek above Gold Basin in the La Sal Mountains, Utah, is an example of a punched laccolith. The small laccolith exposed in the ridge to the left postdates the Mount Peale–Mount Tukuhnikivatz laccolith and bends upward as it approaches the larger, earlier laccolith. View is looking northeast from the north flank of Mount Tukuhnikivatz.

uous series of possible shapes between two distinct end members: punched and Christmas-tree laccoliths (Fig. 8). Gilbert's ideal laccolith (Fig. 1) falls between these two end members.

Punched laccoliths are characterized by small deformation of the overburden beyond the periphery and the development of large-scale shear fractures (slip planes) at, or near, the periphery. The concept was first expressed by Paige (1913, p. 544). Punched laccoliths are associated with elastic-plastic rock behavior, and such rheology is usually only found in the epizone. In the field, punched laccoliths are recognizable by their flat tops, peripheral faults, and steep sides. One example of a punched laccolith is the Mount Peale–Mount Tukuhnikivatz laccolith (Fig. 9) in the La Sal Mountains, Utah. Punched laccoliths are often referred to in the literature as bysmaliths. Unfortunately, the term bysmalith is

also associated with an invalid hypothesis regarding the role of magma viscosity in the mechanical deformation of the roof, and I favor abandoning the term.

Christmas-tree laccoliths are characterized as domes with no peripheral faults, and the beds overlying the intrusion are continuous across the laccolith. If doming has continued far enough, a crestal graben may have formed. Otherwise, the extension over the dome has been accommodated entirely by ductile deformation of the beds. A mechanical model for the formation of Christmas-tree laccoliths within the epizone is presented. However, plastic rheology within the mesozone favors formation of Christmas-tree laccoliths. An excellent example of the smooth dome associated with a Christmas-tree laccolith is Green Mountain near Sundance, Wyoming (Fig. 10). The igneous intrusion is

Figure 10. Green Mountain laccolith near Sundance, Wyoming is an example of a Christmas-tree laccolith. Christmas-tree laccoliths are characterized by lack of a peripheral fault and a rounded dome appearance. The crestal graben in the center of the structure results from radial extension during uplift. The long tree branch just right of center points to the grassy margin on the upthrown block of the crestal graben. The geologic map for this area can be found in Darton (1905). View is looking southeast from crest of Bear Lodge Mountains north of Sundance, Wyoming. The visible dome is 2 km in diameter.

Figure 11. The southern flank of the Christmas Mountains in the Big Bend region of Trans-Pecos Texas. View is looking northeast. The growth of the laccolith has drape folded the Cretaceous carbonate rocks with no faulting. On the left side of the picture, note that the roof is flat where not eroded off. The vertical relief is about 300 m.

not exposed, however, at Green Mountain. The best exposed cross sections of a Christmas-tree laccolith that I have seen are in the Christmas Mountains in the Big Bend region of Texas (Fig. 11) and the Tenmile district (Fig. 12) in Colorado (Emmons, 1898). Jackson and Pollard (1988) present similar cross sections for Mounts Hillers, Holmes, and Ellsworth in the Henry Mountains, Utah. Their profile for Mount Ellsworth is reproduced in Figure 13. In descriptive terms, Christmas-tree laccoliths may be referred to as a Christmas-tree, cedar-tree, composite, or compound laccolith in the literature if a sufficient

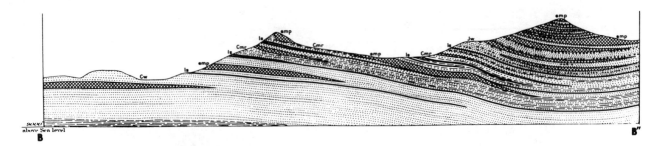

Figure 12. Cross-section B-B″ from Emmons (1898) of a Christmas-tree laccolith in the Tenmile district, Colorado. Jw, Wyoming Formation; ls, limestone in Wyoming Formation; Cmr, Maroon Formation; Cw, Weber Formation; emp, Elk Mountain porphyry.

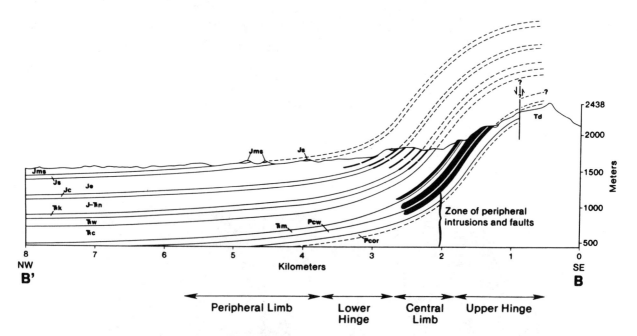

Figure 13. Cross section of Mount Ellsworth in the Henry Mountains, Utah, from Jackson and Pollard (1988). Jms, Salt Wash Sandstone; Js, Summerville Formation; Je, Entrada Sandstone; Jc, Carmel Formation; J-Ṫn, Navajo Sandstone; Ṫk, Kayenta Formation; Ṫw, Wingate Sandstone; Ṫc, Chinle Formation; Ṫm, Moenkopi Formation; Pcw, White Rim Sandstone; Pcor, Organ Rock Shale; Td, diorite porphyry.

cross section is exposed, or as a phacolith if exposure is limited (see Fig. 8).

These end members are presented as an ideal, and perfect examples will seldom be found in the field. The definition of the laccolith in the field is complicated by the present arbitrary depth of erosion. Also, the possible variation between end members is nearly infinite, with the plethora of names for types of laccolithic intrusions in the literature reflecting this variation.

The observed diversity of shapes comes about for at least three reasons:

1. The shape of the laccolith appears to be a function of the way the roof rock is loaded by the invading magma. For an epizonal punched laccolith with a single dominant sill forming the protolaccolith, the loading is essentially on a single surface. For Christmas-tree laccoliths, several load surfaces are present and plastic deformation dominates.

2. Variations in lithology or pre-existing structures may cause a laccolith of intermediate shape, such as a trap-door laccolith, to develop, with the formation of a chonolith (Fig. 4) the extreme example.

3. For mesozonal level intrusions, the rheology of the country rock will be plastic, or possibly visco-plastic, and the deformation will be by plastic flow. Time-dependent rock behavior, however, is not considered in this paper. For a discussion of visco-plastic deformation associated with intrusions, see Ramberg (1981).

CHAPTER II

GENESIS OF LACCOLITHS

As originally suggested by Gilbert (1877), the genesis of laccoliths can be divided into two cycles. The first cycle is the emplacement of the intrusion as a tabular sill or sills. The onset of the second cycle is marked by vertical growth, thickening of the sill(s), and a cessation of radial spreading. The field evidence indicates that the two cycles are mutually exclusive and that radial growth ceases at the onset of vertical growth.

Based on field observation, the emplacement and growth cycles of laccoliths can be further broken down into sequential stages. Since no significant interaction is apparent in the field, each of the stages can be investigated separately. For convenience, I have broken the stages down as follows:

Emplacement. Stage one: the movement of the magma vertically through the lithosphere. Stage two: reorientation of the magma from vertical climb to horizontal spreading.

Growth: Stage three: cessation of horizontal spreading and commencement of thickening. Stage four: large-scale deformation of the overburden by thickening of the intrusion.

ASSUMPTIONS

In order to investigate the problems associated with laccolithic intrusions, I have found it necessary to make several assumptions. My assumptions, and their physical bases, are tabulated below for easy reference:

1. *The level of emplacement is determined by the density contrast between the country rock and the rising magma.* This hypothesis was first proposed by Gilbert (1877), and it is strongly supported by gravity surveys over felsic laccoliths discussed subsequently.

2. *The basic driving force for the magma is a gravitational instability created by the formation of the magma at depth.* If the level of emplacement (assumption 1) is determined by the density contrast between the magma and the country rock, as Gilbert proposed, then the magma pressure must be generated solely by the density contrast between the magma and the lithosphere. Any tectonic magma overpressure should serve to force the magma higher in the lithosphere than the neutrally buoyant elevation. I am unable to eliminate the possibility that tectonic overpressure

in the near surface may be acting on the magmas forming mafic laccoliths, although it seems very unlikely. There is strong evidence, however, that the silicic magmas which form felsic laccoliths do not rise above the neutrally buoyant elevation. There is no field evidence suggesting that mafic laccoliths form differently than felsic laccoliths.

3. *The emplacement of the magma did not significantly affect the physical properties of the country rock.* First, primary hydrothermal alteration is not observed at laccolith margins. Second, the effects of thermal metamorphism due to the intrusion are very small at laccolith margins (MacCarthy, 1925). Laccoliths formed from basic magmas have distinctly more pronounced, but mechanically insignificant, thermal alteration halos due to higher magma temperatures.

4. *The magmas forming the laccoliths in a laccolith group all have similar physical properties at the time of emplacement.* Many authors have remarked on the similarity between the petrography of igneous rocks forming the laccoliths of a given group. Petrographic similarity implies a temperature and chemical similarity of the melts. Thus, it is reasonable to assume that, within resolvable limits, all laccoliths within a group were formed from magmas that were rheologically similar.

5. *The dimensions of the laccolith can be determined from field geological studies.* The unique determination of the density contrast between laccoliths and the country rock by gravity surveying requires that the dimensions of the laccolith be known from field measurements.

6. *The protolaccolith spreads to its final diameter as a thin (<30 m thick) sill.* The evidence for spreading to the full diameter as a thin sill is summarized in Figures 14, 15, and 16. As Davis (1925) first pointed out, if the laccolith grows radially only, as shown in Figure 14, the invaded rocks must be rotated as distinct blocks, often in excess of 30°, and then flattened out again, since the roofs of most large laccoliths are presently relatively flat. Multiple block rotations would certainly result in a trail of radial hinge zones behind the advancing intrusion. A laccolith that grew both radially and vertically would also leave a trail of radial hinge zones after time t_2 (Fig. 15) and would be indistinguishable from model 1 (Fig. 14) after time t_2. These remanent hinge zones have

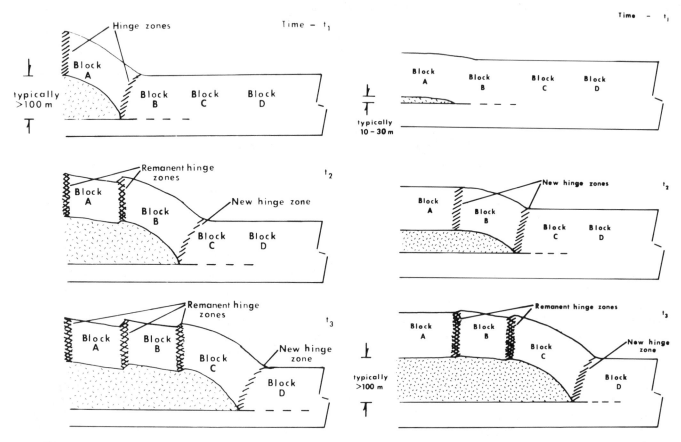

Figure 14. Possible modes of laccolith growth, model 1. Radial growth only, thickness remains constant. This is model B of Hunt and others (1953, Fig. 70, p. 142). The remanent hinge zones implied by the model have not been observed.

Figure 15. Possible modes of laccolith growth, model 2. Simultaneous radial and vertical growth. After time t_2, model 2 is indistinguishable from model 1, and the arguments against model 1 apply equally to this model. This is model C of Hunt and others (1953, Fig. 70, p. 142). The plate-bending models of Pollard and Johnson (1973) and Jackson and Pollard (1988) would produce a sequence of intrusion similar to this model. Again, no trail of remanent hinge zones has been found.

not been observed in the field or reported in the literature. It is reasonable to assume that, if they existed, these radial hinge zones would be preserved and visible on at least some large laccoliths.

If a large laccolith were bulldozing its way through the country rock, as shown in Figure 14, the margin must be deformed. The flank of the laccolith forming Mount Peale–Mount Tukuhnikivatz in the La Sal Mountains of Utah is exceptionally well exposed in a glaciated canyon and, as can be seen in Figure 17, the beds adjacent to the laccolith are virtually undeformed. Deformation is limited to within 3 m of the laccolith margin. Thus, this laccolith was not advancing radially at its full thickness. Numerous other examples of this lack of horizontal deformation at the laccolith margin exist. The evidence for model 3 (Fig. 16) appears conclusive. Therefore, the laccolith must grow to its full diameter as a thin sill.

7. The overburden acts as a continuum. Overburden behav-

ior as a continuum was demonstrated in Australia. As reported by Brawner (1974, p. 763–765) at the Latrobe Valley open-pit coal mine in Victoria, the floor is underlain by an artesian aquifer at a depth of 76 m. The floor of the pilot opening was 91 m wide and did not heave. When the floor was widened to 107 m, the floor heaved up to 3.7 m. The "laccolith effect" interfered with mining operations, so the induced experiment in large-scale rock deformation was stopped by dewatering the aquifer. The large-scale experiment at Latrobe Valley illustrates the ability of shallow, unconfined rocks to act as a continuum.

If the shallow crust acts as a continuum, then theoretical analysis treating the overburden as a plate is valid. In addition, inelastic deflection of the overburden does not occur until the total load exceeds the yield strength of the overburden. These are crucial requirements if the protolaccolith is to spread to its full diameter as a thin sill.

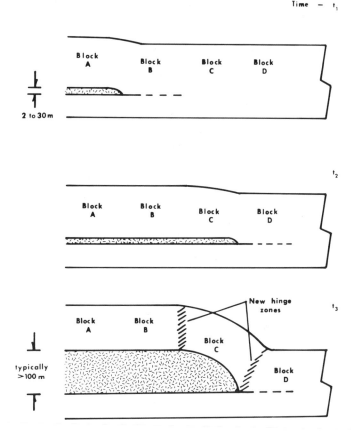

Figure 16. Possible modes of laccolith growth, Model 3. Radial growth only until time t_2, followed by dominant vertical growth after t_2. This model best fits the field evidence and is similar to model A of Hunt and others (1953, Fig. 70, p. 142). However, no evidence has been found that the laccolith is fed laterally by a central stock.

FEEDER DIKES

Hunt's model. Lateral feeders from a central stock

Several authors, most notably Hunt and others (1953) and Hunt (1980), have proposed that laccoliths are fed by lateral, tonguelike feeders from a central stock. In the Henry Mountains, Utah, and elsewhere, the laccoliths tend to form clusters, with a large laccolith, or possible stock, located approximately in the center of the cluster. Hunt and others (1953) and Hunt (1980) envision the large central body as sending out lateral shoots, which formed the smaller laccoliths surrounding the central body (Fig. 18). It is obvious, however, that many laccoliths have no relation to any other laccolith and that, commonly, laccoliths stand many kilometers distant from any other igneous intrusion.

The most distant laccolith from a central cluster in the Henry Mountains is Trachyte Mesa, about 10 km from Mount Hillers. The magnetic data of Case and Joesting (1972) do not indicate a connection by an igneous body between Mount Hillers and Trachyte Mesa in the subsurface. Further, no evidence for a lateral connection between these laccoliths has been found exposed on the surface. The gravity data, as discussed subsequently, do not indicate central stocks, as proposed by Hunt and others (1953) or Hunt (1980), in the Henry Mountains, except at Mount Ellen. Jackson and Pollard (1988) have recently mapped the southern Henry Mountains in detail, and find no evidence to support Hunt's hypothesis.

In arguing for a stock acting as a lateral feeder (Fig. 18) for the laccoliths in the Henry Mountains, Hunt and others (1953) make two key observations. One is that the cores of the principal peaks are occupied by stocks (i.e., discordant intrusions), as illustrated by Figure 18. Secondly, most of the smaller laccoliths around these central peaks are elliptical, with their semimajor axes aligned normal to the central peak.

For the central stock to bow up the flanking rocks, the shear stress, τ_{rz}, on the magma–country rock interface must be greater than the body forces, σ_z, holding the rocks on the flanks down (Fig. 18). As shown below, the shear stress at the magma–country rock interface is effectively zero. Hence, the magma cannot be lifting the surrounding rocks in shear, as Hunt's model (Fig. 18) implies. Hunt's model also requires that the central peaks be the same age as the smaller laccoliths that surround them, or how could the stocks within the central peaks act as feeders? These age relations have not been tested in the Henry Mountains or elsewhere.

Alternatively, if the central peaks were in existence, then later dikes will tend to be radially distributed around them as at Spanish Peaks, Colorado, and elsewhere (see Hyndman and Alt, 1983, 1987; and Muller, 1986 for most recent work). If these radially distributed dikes were acting as feeder dikes for the later laccoliths, rather than the central stock proposed by Hunt and others (1953), then their radial distribution is a necessary consequence. A small laccolith formed above a feeder dike will tend, all other factors being equal, to have its long axis aligned with the underlying feeder dike. The axial alignment is thus also easily explained. Further evidence against a stock feeder is evident at Trachyte Mesa in the Henry Mountains, Utah. The sill exposed on the northwest flank was advancing in a direction perpendicular to the semimajor axis, not away from Mount Hillers as proposed by Hunt and others (1953).

The last item to be resolved in Hunt and others (1953) model is the supposed presence of discordant intrusions in the core of the central peaks. Recent detailed mapping by Jackson and Pollard (1988) shows that the intrusions in the central peaks are generally concordant (Fig. 13), not discordant. Any model proposed for the formation of these peaks must be capable of lifting the surrounding country rock. The roof will be uplifted if a laccolith is first intruded beneath them as shown in Figure 19. As the laccolith grows, the crest is placed in extension and a crestal graben will eventually form (Fig. 19). As growth continues, this crestal graben may founder in the magma, or magma may exude or be squeezed up by the foundering crestal block along fractures

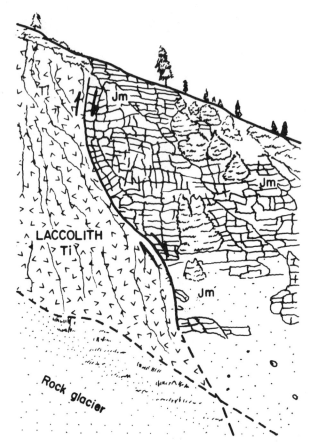

Figure 17. Sheared margin of Mount Peale–Mount Tukuhnikivatz laccolith in the La Sal Mountains, Utah. View is looking south from the head of Brumley Creek. The laccolith (Ti) is the massive rock on the upper left. The layered sedimentary rock on the upper right is Morrison Formation (Jm) largely covered by talus. The regional dip is to the left and postdates the emplacement of the laccolith. An overview of this area is given by Hunt (1958, his Fig. 118).

in this zone of weakness. The end result is that the crestal area of the roof may be filled from the magma forming the laccolith (or later intrusions), which is discordant with the country rock. Hence, the intrusion appears to form a stock as shown in Figure 19. An example of an eroded neck probably formed in this fashion is shown in Figure 20. The model shown in Figure 19 satisfies the field relations proposed by Hunt and others (1953) observed in the Henry Mountains, Utah, without the mechanical drawbacks of their model. Further, the model in Figure 19 is consistent with other observations (e.g., Corry and others, 1988; Jackson and Pollard, 1988). My model appears to be quite similar to that proposed by Hyndman and Alt (1983, 1987) for the Adel and Highwood Mountains in Montana.

Feeder dikes below the laccolith

There is abundant evidence that laccoliths are fed from below by feeder dikes. A possible feeder dike was first reported by Harker (1904) on the Isle of Skye. Well-exposed feeder dikes were first reported by Gould (1926a) for the Mount Mellenthin laccolith in the La Sal Mountains, Utah. The feeder dikes are exposed in a remote glacial valley at the head of Horse Creek. Though Hunt's (1958) criticism that the relations are obscured by talus is in part justified, one of the three feeder dikes reported by Gould (1926a) can be seen to definitely merge with the Mount Mellenthin laccolith, and the others probably do. Feeder dikes for a separate laccolith exposed in the ridge between Mount Mellenthin and Mount Peale were also found (Fig. 21). The feeder dike is exposed nearly continuously for about 200 m vertically in the headwall of a cirque in the north fork of Brumley Creek.

Feeder dikes have also been reported by Chadwick (1944) in Acadia National Park, Maine; by Lyons (1944) in the Big Belt Mountains, Montana; and by Larsen and Cross (1956) in the San Juan Mountains, Colorado.

Readily observable feeder dikes are exposed at Bee Mountain laccolith in Texas. The feeder dikes at Bee Mountain are visible from Texas Highway 118 about 2 km north of Study Butte, Texas (Figs. 22 and 23).

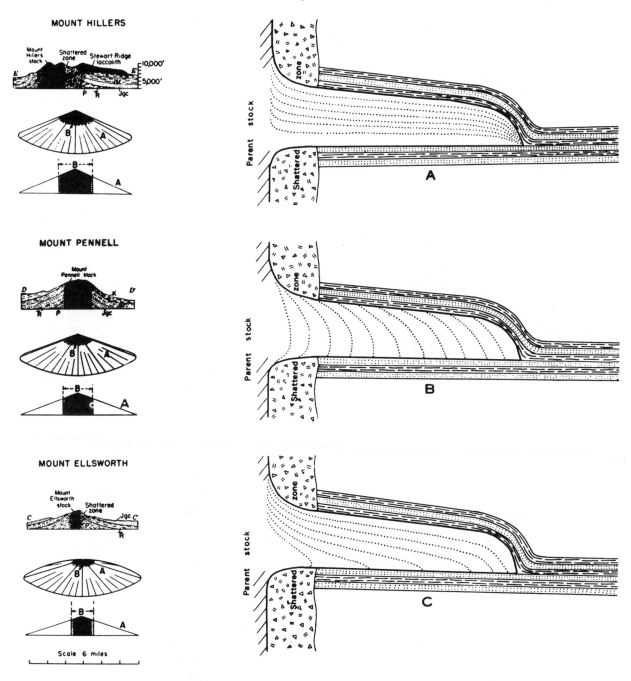

Figure 18. The diagrammatic cross sections on the right illustrate some alternative ways by which the Henry Mountains laccoliths may have grown from a central stock as suggested by Hunt and others (1953). The dotted lines represent stages during growth. The laccoliths may have developed from sills that lifted their roofs (A), or they may have grown laterally at their full thickness (B), or they may have grown by spreading and thickening (C). Hunt and others (1953) favor model C. The field evidence and recent work by Jackson and Pollard (1988) do not support any of these models. Hunt and others' (1953) model also requires that the parent stock and the attached laccoliths be of the same age.

The doming associated with the central stocks postulated by Hunt and others (1953) is shown on the left. The body forces, σ_z, act vertically downward on the country rock. To lift the country rock in the manner shown requires that the shear stress, τ_{rz}, on the boundary of the magma (shown in black) with the country rock exceed the body forces acting on the country rock. As will be shown, the shear stress, τ_{rz}, between the country rock and the magma is effectively zero. Therefore, a central stock would not lift the country rock in the manner proposed by Hunt and others (1953) but, as shown by Jackson and Pollard (1988), would laterally compress the country rock.

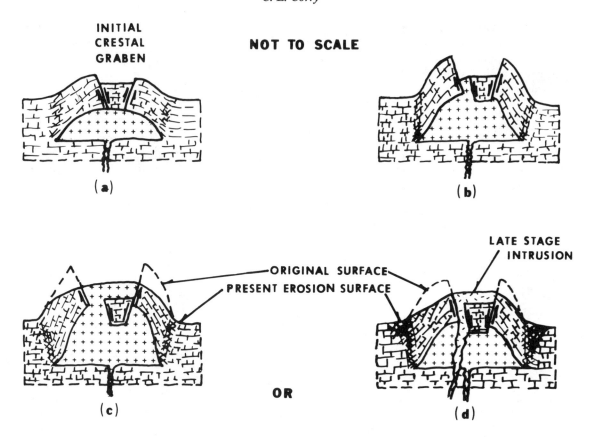

Figure 19. The sequential formation of a laccolith with a crestal graben that founders and is subsequently intruded by magma from the laccolith or a later intrusion. a. Laccolith grows to point where crestal graben forms. b. Continued growth pulls country rock further apart, and crestal graben block begins to founder. c. Continued growth may completely founder the crestal graben block, and magma from the laccolith may extrude up and fill the central basin; or d. Laccolith growth stops at stage b, but laccolith is cut by later intrusion which fills the central basin and partially or completely covers the original graben block.

The end result may look like a stock at the surface but it is still a floored intrusion and classified as a laccolith. An example of a foundered crestal graben block can be found in the Solitario, Trans-Pecos Texas (Corry and others, 1988).

McBride (1979) and Hyndman and Alt (1982, 1983, 1987) have reported on feeder dikes in the Big Belt Mountains, Montana. While I have not had an opportunity to examine the relationships in the field, the descriptions by McBride (1979) suggest to me that the small laccoliths(?) (I would prefer chonolith for features such as Lionhead, Birdtail, Haystack, and Fishback) in the Big Belt Mountains may be a higher level exposure of plugs and associated dikes similar to those described by Delany and Pollard (1981) at Ship Rock, New Mexico.

CONTROL OF THE LEVEL OF INTRUSION

Gilbert's hypothesis

In attempting a quantitative analysis in structural geology, it is difficult to determine the boundary conditions at depth. Direct observation is impossible, and any information must be obtained by inference, usually long after the event.

Gilbert's hypothesis, quoted above, that laccoliths are emplaced at the neutrally buoyant elevation, would, if valid, simplify the problem of establishing boundary conditions for laccolith emplacement. If the dominant factor controlling the level of emplacement is density contrast between the rising magma and the weighted mean density of the country rock, then regional stresses, magma rheology and type, lithology of the country rock at the level of emplacement, and mechanical properties of the overburden can be neglected for the lower boundary conditions.

Change in magma density as it solidifies

Field evidence. A major difficulty in substantiating Gilbert's hypothesis is the unknown amount the magma increases in density as it cools and crystallizes. Gilbert (1877, p. 74) and

Figure 20. The volcanic neck forming Devils Tower in the Black Hills of Wyoming may have formed due to roof failure in the central zone of extension above the upper level of a Christmas-tree laccolith. While not usually so spectacular, similar jointing is common in laccoliths. A similar neck formed as a late-stage feature at Syowa Sinzan, Japan (shown subsequently).

Avakian (1970) considered this problem for the Henry Mountains of Utah. They obtained a weighted mean density for the rocks of ~2,340 kg m^{-3}, and a density for the intrusions of 2,610 kg m^{-3}, or a density contrast $\simeq$270 kg m^{-3}. From this, Gilbert inferred that the total increase in density of the magma during solidification was about 10 percent. In gravity surveys, a density contrast of 270 kg m^{-3} would yield a readily identifiable anomaly for bodies the size of the major laccoliths in the Henry Mountains (see Appendix B). From the gravity survey of the Henry Mountains shown on Plate I, it can be seen that no anomalies reflecting a density contrast as large as 270 kg m^{-3}, with wavelengths corresponding to the dimensions of the laccoliths, are identifiable. In fact, the gravity survey of the Henry Mountains (Plate I, and as discussed below) indicates that the laccoliths do not have any resolvable density contrast with the surrounding country rock. The density measurements of Gilbert (1877) and Avakian (1970)

must either overestimate the bulk density of the intrusions, or underestimate the weighted mean density of the country rocks, or both.

Gilbert (1877, p. 71) "... tends to sustain the view that the laccolitic rocks contracted less in cooling than the volcanic. The prismatic structure is produced by the contraction of cooling rocks during and after solidification. That this does not occur in the Henry Mountain trachytes indicates that their contraction was small."

Many of the magmas that form laccoliths are a crystalline mush at the time of emplacement, as evidenced by the large percentage of phenocrysts in the glassy margins of the intrusion. In the magma chamber forming the laccolith, any effect of crystallization on density increase will be opposed by the fact that the more dense crystals formed before emplacement, as noted by Hunt and others (1953). Therefore, only the density of the liquid part of the magma will change significantly during cooling. I estimate from observations of chilled zones at felsic laccolith margins that many silicic magmas were $\geqslant$25 percent crystalline at the time of emplacement.

Experimental work. Experimental attempts to determine the change in density between a crystalline rock and a melt of the same composition are limited. The early work is summarized in Daly and others (1966) and is based largely on the work of Douglas (1907). Much of the experimental work is based on the fact that glass is a supercooled liquid. Hence, the density of the glass should closely represent the density of the fluid, or melt. By first measuring the density of the whole rock, then melting it and quenching the melt to obtain a glass, and then determining the density of the glass, an approximation of the density variation between a crystalline solid and a melt can be obtained. Working with samples "about the size of an acorn" (~1 cm^3?), Douglas (1907) attempted the above procedure. From the description of his experimental technique, any volatiles in the rock must be driven off during melting, and he probably underestimates the density change. An additional shortcoming is that no estimate of crystal size versus sample size is given for crystalline rocks used in his study. Thus, a mineral of a given density may be over-, or under-represented in his samples relative to the bulk composition of the rock. Despite these shortcomings, the variation in density found by Douglas (1907) agrees quite well with the more recent work by Murase and McBirney (1973). Murase and McBirney (1973) measured the variation of density with temperature directly for the same rock types; their work is summarized in Figure 24.

The experimental work by Douglas (1907) and Murase and McBirney (1973) suggests that for a felsic rock, such as rhyolite, the increase in density from a molten state to a solid will be approximately 100 kg m^{-3}. The work of Murase and McBirney (1973) also suggests that the variation in density during solidification of the magma will increase as the rock becomes more mafic (Fig. 24), to about 180 kg m^{-3}. The data given by Douglas (1907) for mafic rocks is more abundant and, hence, more scattered, but an increase of ~180 kg m^{-3} is a reasonable average.

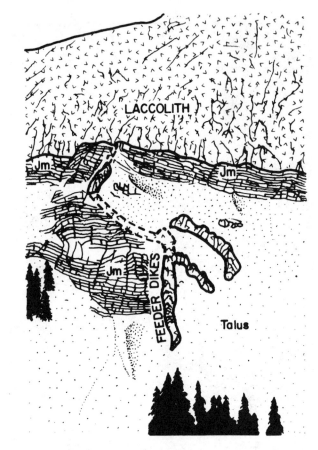

Figure 21. Feeder dikes exposed in the headwall of a cirque at the head of the north fork of Brumley Creek, La Sal Mountains, Utah. View is looking east. The laccolith forms a ridge between Mount Mellenthin and Mount Peale. The feeder dike is exposed in the headwall for about 200 m.

Experimentally, only one sample, a basalt (GOB), of those tested by Murase and McBirney (1973, their Fig. 10) showed an inflection in the curve that they attributed to crystallization. After crystallizing, the basalt showed very little change in density with further cooling, and the total change in density from a liquid to a crystalline solid was 150 kg m^{-3}. Douglas (1907) included two crystalline granites in his experiments. For these two samples he obtained density variations of 210 kg m^{-3} and 254 kg m^{-3}. Only three of the fifteen experiments Douglas (1907) ran, including the two crystalline granites, have density variations of >200 kg m^{-3}. All three of the samples with density variations >200 kg m^{-3} were noted to have experimental difficulties, and Douglas had to crush and remelt the glass of his original melt at least one additional time for all three of these samples. Thus, experimental error may adversely affect the results of these samples. If these suspect values are excluded, the results from Douglas (1907) indicate a mean increase in density for all samples of 140 kg m^{-3}. The data from Murase and McBirney (1973) indicate a mean density increase during solidification of 130 kg m^{-3}. The data from both sets of experiments also indicate that density increase of silica

magma during cooling will be on the order of 100 kg m^{-3}; for a basic magma, an increase of ~180 kg m^{-3} is suggested.

Theoretical work. Huppert and Sparks (1984) review the work on the effects of crystallization on the change in density during cooling from a liquid to subsolidus temperatures. The theoretical work presently contains too many unknowns to adequately define the density variation and is mainly concerned with basic magmas.

Summary. While there is no definitive information on the increase in density during solidification, the available evidence suggests that an increase in density on the order of 100 kg m^{-3} for silicic magmas, to 200 kg m^{-3} for basic magmas, or more for ultrabasic magmas, can be expected. Since basic magmas usually do not contain as many phenocrysts, a larger percentage of the magma will change in density than in a more porphyritic, felsic intrusion. Hence, it is reasonable to expect mafic intrusions to exhibit a larger density contrast than their felsic cousins.

Neutrally buoyant elevation

The *neutrally buoyant elevation* is the elevation at which the

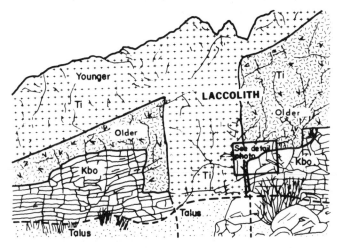

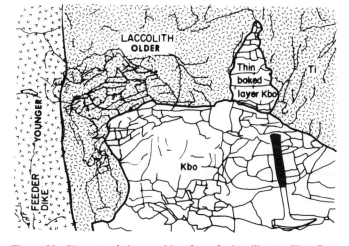

Figure 22. Feeder dike for Bee Mountain laccolith, 2 km north of Study Butte on Texas 118 (see Yates and Thompson, 1959). Kbo, Cretaceous Boquillas Formation; Ti, Tertiary laccolith. The feeder dike shows evidence in outcrop of multiple intrusions through the channel. The overall width of the dike is 4 m. Plastic deformation of the country rock is evident in outcrop on the right margin of the feeder dike. The transition from vertical movement to horizontal in the outlined area is shown in detail in Fig. 23.

Figure 23. Close-up of the transition from feeder dike to sill at Bee Mountain laccolith near Study Butte, Texas. An overview of the feeder dike is shown in Fig. 22. Kbo is Cretaceous Boquillas Formation.

magma will have the same density as the weighted mean density of the country rock. Only very near the surface will the weighted mean density be approximately equal to the country rock density at the level of intrusion. The weighted mean density becomes progressively less than the country rock density as depth increases. For the hypothetical crustal model shown in Figure 25, the weighted mean density is 100 kg m^{-3} less than the local country rock density at a depth of 3 km. The expected increase in density during solidification of a silicic magma is ~100 kg m^{-3}. Hence, the present density contrast, $\Delta\rho$, between a felsic laccolith emplaced at the neutrally buoyant elevation and the country rock is near zero. Conversely, any other factor that may affect the level

of emplacement of felsic laccoliths will result in a significant density contrast unless, fortuitously, the other factors always cause emplacement within the vertical range of neutral buoyancy.

The contrast between local country rock density and the weighted mean density increases with depth (Fig. 25). Thus, silicic magma intruding at depth has a substantial negative density contrast with regard to the local country rock at the neutrally buoyant elevation. Even after solidification, a deep felsic intrusion may have a negative density contrast of about 100 kg m^{-3}.

For a basic magma, which may increase in density by 150 to 200 kg m^{-3} (Fig. 24) during solidification, the final density contrast, $\Delta\rho$, will be positive for a mafic laccolith emplaced within the shallow crust at the neutrally buoyant elevation. A positive $\Delta\rho$ is borne out by gravity surveys at Mustang Hill, Texas

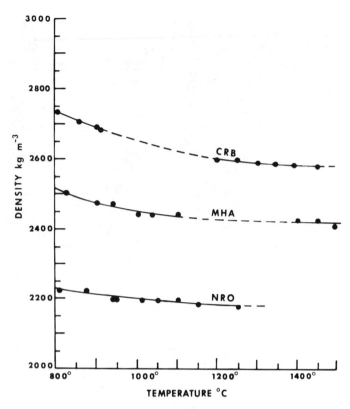

Figure 24. The variation of density as a function of temperature for three types of igneous rocks: Columbia River basalt (CRB), Mount Hood andesite (MHA), and Newberry rhyolite (NRO). Data from Murase and McBirney (1973). Room-temperature densities of samples quenched from liquids at 1,400°C are CRB = 2,760 kg m^{-3}, MHA = 2,590 kg m^{-3}, and NRO = 2,280 kg m^{-3}.

(Greenwood and Lynch, 1959), Trompsburg, South Africa (Buchmann, 1960), Mara, Tanzania (Darracott, 1974), and the Channel Islands, Great Britain (Briden and others, 1982). Deep mafic intrusions should have near zero density contrast after solidification if the estimates of density increase during solidification are correct.

The tacit assumption is made that the weighted mean density within the crust increases as a relatively smooth function, as illustrated in Figure 25. Hence, the level of neutral buoyancy has a finite vertical range, and the magma of the protolaccolith is emplaced in country rock of higher density. This may not always be the case, the extreme example being a magma that rises to the air-rock interface. An equivalent interface may occur within the earth, for example at the basement (with a density of, say, 2,800 kg m^{-3}) and sedimentary rock (density of, say, 2,400 kg m^{-3}) interface. A magma with a density of 2,600 kg m^{-3} rising through the dense basement will spread laterally at the basement-sedimentary rock interface, which is the neutrally buoyant elevation, but will grow into rocks of lesser density. After solidifying, the laccolith will be out of density equilibrium and will possess a

significant density contrast with respect to the surrounding low-density rocks. This phenomena may occur any time the thickness of the laccolith is substantially greater than the vertical range of the neutrally buoyant elevation. In defense of the arguments that follow, I have never encountered the hypothetical situation outlined above. It is, however, geologically realistic to expect that some felsic laccoliths possess a positive density contrast, for the reasons stated above, yet were emplaced at their neutrally buoyant elevation.

Another factor that may affect the apparent density contrast, and also the final thickness of the laccolith, is the possible loss of the roof rock during growth, either by foundering in the magma forming the laccolith, or shedding by lateral gravity sliding. As discussed subsequently, the loss of the roof during growth may allow the laccolith to become very thick and possibly grow into country rock of lower density. Such laccoliths would be geophysically indistinguishable from a stock since they would have a distinct positive gravity anomaly.

Vapor barriers

A review of factors that may control the level of emplacement has been given by Mudge (1968). He finds three factors that may control depth of intrusion: (1) the presence of a well-defined parting surface; (2) the effects of overburden or lithostatic pressure; and (3) the presence of a "vapor barrier."

The effect of a well-defined parting surface is discussed further in stage two of the emplacement process. The second factor, the effect of overburden, or lithostatic, pressure is reexamined with punched model three in the section on theoretical models.

The third factor proposed by Mudge (1968, p. 328) is that "A ductile layer may form a physical barrier to retard upward advance of the magma, but also it may have served as a vapor barrier, exerting control by confining vapor pressure." The presence of a vapor barrier would cause the magma to be emplaced below its neutrally buoyant elevation, with a resultant negative density contrast for a felsic laccolith. As concerns laccolithic emplacement, it is difficult to accept Mudge's hypothesis for a vapor barrier that appears to require ". . . the role of steam and superheated vapor pressure may be . . . an important factor in assisting the magma to emplace concordant bodies" (Mudge, 1968, p. 326). A loss of volatiles from the magma, or heating the ground water by igneous intrusion, leaves distinct traces in the country rock, both at the contact with the intrusion, and at considerable distances from the intrusion. The lack of primary hydrothermal alteration of the country rock at laccolith margins is well established.

Stress barriers

A mechanically valid hypothesis for the control of the level of emplacement of mafic sills was presented by Gretener (1969). Based on a mechanism first advanced by Anderson (1951), he proposed that some beds within the shallow crust act as "stress

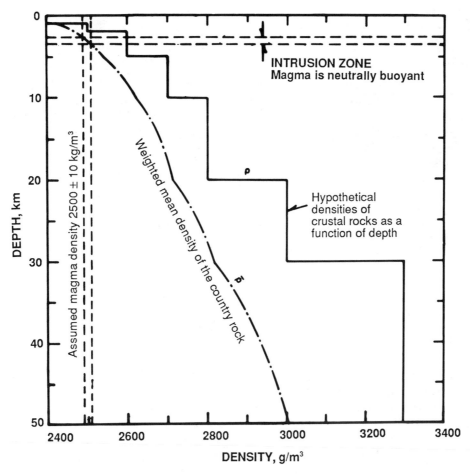

Figure 25. Weighted mean density as a function of depth for an assumed range of densities within the lithosphere. The intrusion zone shown assumes a silicic magma with a liquid density of 2,500 ± 10 kg m^{-3}. The magma is assumed to be incompressible and to rise along the adiabat for reasons discussed in the text. At the time of emplacement the magma has a negative density contrast relative to the local country rock of $\Delta\rho$ = –100 kg m^{-3} at the neutrally buoyant elevation. During solidification the density of the intrusion increases by ~100 kg m^{-3}. Thus, the final density contrast at the level of intrusion is approximately zero. Silicic magma emplaced at greater depth would have a negative density contrast.

barriers" to the vertical advance of basic magmas. Such stress barriers would be formed by beds under a differential stress. The differential stress may arise from a regional stress field or from the pressure on the walls of the intruding dike. Engelder and Sbar (1984) have noted that near-surface horizontal stresses may commonly exceed the vertical body force stresses. Under horizontal compression, the stronger beds within the crust, such as dolomites and quartzites, will bear a disproportionate share of the load and act as "stress concentrators." These preloaded beds will resist fracturing by the advancing dike, and sills will therefore tend to form beneath them. However, in laccolithic areas, sills commonly form near the top of the competent beds instead of at their base. After a field examination in the Henry Mountains, Utah, Gretener (personal communication, 1974) conceded that no evidence for a stress barrier could be found.

The mechanical validity of a stress barrier is unquestioned but no evidence for the existence of a stress barrier at the time of emplacement has yet been found in laccolithic areas. The presence of a stress barrier would also cause the magma to be emplaced beneath its neutrally buoyant elevation, with a resultant negative density contrast for a felsic laccolith. No laccolith with a negative density contrast has yet been found.

Stopping of rising magma at a freely slipping crack

Weertman (1980) has provided a quantitative model for stopping a rising, liquid-filled crack at the intersection with a horizontal, freely slipping joint. If the rock is under horizontal compression, he shows that the freely slipping condition can be relaxed. Engelder and Sbar (1984) have shown that large hori-

Figure 26. Diagram to illustrate a hypothetical explanation of flat-topped dikes. After Gilbert (1877). While a complete cross section is not exposed in the field as shown, the available outcrop, and the description by Gilbert, are convincing evidence for his diagram. Weertman (1980) quantified this behavior in a model for stopping a rising, liquid-filled crack at the intersection with a horizontal, freely slipping joint. The reorientation to horizontal spreading also reorients the local stress field above the sill, and vertical cracks become freely slipping joints. The magma then reorients to vertical climb again at the next vertical joint unless it is at its neutrally buoyant elevation.

zontal compressive stresses are likely. Hence, Weertman's model should be valid.

Vertical cracks are probably as common as horizontal cracks in the subsurface. Under the horizontal compression, which favors reorientation of the magma from vertical climb to horizontal spreading, vertical cracks would tend to be closed. However, once the magma has reoriented to horizontal spreading, it exerts a vertical stress on the country rock. The vertical stress exerted by the magma would cause any nearby vertical cracks to act as freely slipping joints. Hence, any magma that is migrating horizontally below its neutrally buoyant elevation will reorient to vertical climb at the first suitable vertical joint. The reason for reorientation to vertical climb is the same as Weertman (1980) gives for reorientation to horizontal spreading, namely the presence of a freely slipping crack perpendicular to the plane of intrusion. At the neutrally buoyant elevation, the magma would continue to migrate horizontally even though a vertical crack might be present, since the vertical climb force is zero.

Gilbert (1877, p. 28–29) first described flat-topped dikes that behaved in the fashion quantified by Weertman (1980) at Mount Holmes in the Henry Mountains, Utah. Gilbert (1877, p. 29) also noted that while the rising magma stops and spreads laterally at a horizontal joint, upward migration of the magma would continue after some offset, as shown in Figure 26. Thus, no effective barrier to vertical migration exists.

Further evidence against any barrier to vertical magma migration is the presence of extrusive igneous rocks in time and spatial proximity with the laccoliths. If any barrier to vertical migration existed, it should have been a barrier to all magmas, not just some.

All presently known mechanisms for magma transport through the asthenosphere and lithosphere depend primarily on magma buoyancy for vertical movement. The proposed mechanisms for stopping the vertical migration of magma all postulate the magma being held up below the neutrally buoyant elevation. Though a mechanism, such as tectonic overpressure, might conceivably move the magma above the neutrally buoyant elevation, I know of no geologically realistic mechanical model that accounts for such transport.

CHAPTER III

GRAVITY AND MAGNETIC SURVEYS OF LACCOLITHS

TESTING GILBERT'S HYPOTHESIS FOR DENSITY CONTROL OF EMPLACEMENT LEVEL

Criteria

Use of gravity surveys to determine density contrast. Since the dimensions can be independently determined, a gravity survey of a laccolith can be used to determine uniquely the present density contrast between the intrusion and the surrounding country rock. As shown in the definition of neutrally buoyant elevation, a felsic laccolith emplaced high in the crust at the neutrally buoyant elevation should have near zero density contrast after solidification. A mafic laccolith emplaced at the neutrally buoyant elevation high in the crust should have a positive density contrast of approximately 100 kg m^{-3} after solidifying.

During the course of this investigation, approximately 3,000 gravity stations were occupied, though not all data are presented here. My gravity data are on file with the Defense Mapping Agency (DMA) Gravity Library at the St. Louis Aerospace Center. Theoretical gravity was computed using the 1967 International Geodetic Formula. All stations were tied to the gravity base-station network maintained by the DMA Gravity Library. The density used in the Bouguer correction is the standard value of 2,670 kg m^{-3} in all surveys.

Signal-to-noise ratio. In order to determine whether a given body can be detected by a gravity survey, it is necessary to first make an estimate of the errors occurring in the field acquisition of the data and in the corrections applied in data reduction (i.e., the noise level).

One of the largest sources of errors in gravity surveys of laccoliths are the terrain corrections. Because large laccoliths form high, rugged mountains, all gravity surveys over them must be reduced to complete Bouguer anomaly maps. In my surveys, the Hayford-Bowie scheme for terrain corrections (Swick, 1942) was used, and the corrections were extended to 167 km (zone O) from the gravity station. An estimate of the error introduced by the terrain correction procedure can be made from the survey of the Henry Mountains (Plate I). The terrain corrections range up to 57 mGal on Mount Pennell, while in the lowlands terrain correc-

tions were only 5 to 10 mGal. Along line A–A″ on Plate I, the vertical range in gravity station elevations is 3 km, and the range of terrain corrections is from 5 to 57 mGal. Since the mountains are rugged, the variation in terrain correction between closely spaced stations is great. Thus, any random errors in the terrain corrections will be evidenced as high spatial frequency noise in the contours. As is evident in the map (Plate I), the contours are smooth across the mountains even though the terrain correction has a range of more than 50 mGal. The largest variation is observed around Mount Pennell where terrain corrections are greatest. However, the diorite porphyry of the Mount Pennell laccolith is cut by a monzonite porphyry (Hunt and others, 1953), and the anomalies observed in the profile A–A′–A″ may be real rather than "noise" from terrain correction errors. Nearly as great a variation in terrain correction was found near Mount Ellen without any "noise" in the data. Since the contour interval is 2 mGal, "noise" with amplitude greater than ±0.5 mGal should be obvious. It appears that, in general, the errors associated with the terrain corrections are ≤±0.5 mGal.

Significant errors in gravity data result from inaccuracies in the elevation used for the free-air and simple Bouguer calculations. Following standard field practice for determining elevations, two or more aneroid altimeters were used in conjunction with the gravimeter. The elevation of the stations was also estimated from the contour maps, usually 7½-minute quadrangles. Stations at points of known elevation, such as bench marks, mountain tops, and road junctions, were occupied as frequently as possible during each field day. The altimeters were drift corrected as a function of time between points of known elevation. The drift-corrected altimeter readings were then compared with the elevation estimated from the contour map. In most cases good agreement was found between the map elevation and the altimeters. Occasionally it was necessary to adjust the station position, usually up or down a stream bed, to obtain agreement between map and altimeter elevations. The final elevations used are estimated to be accurate to better than ±3 m, with a worst-case accuracy of ±5 m. This is equivalent to a worst-case error of ±1.0 mGal, and an average error of <±0.6 mGal.

Another possible source of error is drift in the gravimeter. Most gravity stations were taken using LaCoste and Romberg gravimeters. The meter used for most of my surveys (G135) showed an instrumental drift of <1 mGal in three years. In addition to instrument drift, earth tide effects can reach an amplitude of ±0.3 mGals. To compensate for earth tides and instrument drift, a base station was occupied in the morning and evening during all surveys, and occasionally during the course of the day. Since earth tides have a period of approximately 12 hours, the base station readings can be used to calculate a drift correction and earth tide correction for the gravimeter. Stations taken at the same point on different days normally repeat to better than ±0.2 mGal. Since 7½-minute quadrangles were used almost exclusively for the gravity surveys, the latitude error is negligible. Thus, the worst-case error for an individual station is ±1.7 mGal in my surveys. Since random errors of this magnitude would be evident as noise of high spatial frequency on the contour maps, and such noise is not apparent, I conclude that the stations are generally accurate to approximately ±0.5 mGal, or a noise level of 1 mGal.

It is a precept of communication theory that the smallest signal that can be unambiguously resolved has an amplitude of at least twice the noise level (i.e., the minimum signal-to-noise ratio is 2:1). The amplitude of the smallest anomaly that could possibly be detected is then 2 mGal for the estimated noise level of 1 mGal.

The question of minimum resolution of the gravity survey is important to the method, that is, what minimum density contrast of some small body can be detected? Using a Bouguer slab approximation, a laccolith 0.25 km thick, with a density contrast of 100 kg m^{-3}, will produce a gravity anomaly of 1 mGal. Since a 2 mGal anomaly is the approximate lower limit of a resolution, laccoliths must be either >0.25 km thick, or have a density contrast >100 kg m^{-3}, or both, if they are to be detected. However, a laccolith with zero density contrast could be infinitely thick and not be resolved by a gravity survey.

Spatial wavelengths in gravity data. An important point in the analysis of the gravity data with relation to laccoliths is the concept of spatial wavelengths of anomalies. Theoretical solutions exist for a variety of regularly shaped bodies (e.g., cylinders, spheres, etc.). From these solutions, it has been deduced that for a body of given dimensions, at a given depth, only anomalies of a certain spatial wavelength are generated. For bodies that lie at the surface, such as exposed laccoliths, the wavelength, λ, of the anomaly is directly related to the size of the body causing the anomaly. For example, a laccolith with a diameter of 6 km would have a gravity anomaly with a wavelength of 6 to 8 km. Thus, an anomaly of much greater, or lesser, wavelength is related to a larger, or smaller, body than the laccolith. Laccoliths are commonly cut by later intrusions (e.g., Mount Pennell). These later, smaller intrusions are sometimes seen in the gravity field as high-

frequency (i.e., short spatial wavelength) "noise" or small intrusions may cause aliasing, as discussed below.

Another problem in the analysis of the gravity data is that many laccoliths appear to be emplaced over deep-seated massive intrusions that perturb the gravity field at longer wavelengths. Therefore, it is necessary to eliminate any anomaly with a wavelength differing significantly from the dimensions of the laccolith. The process is known as removing the regional field to obtain a residual anomaly.

The terms *high frequency* or *low frequency* are commonly used when describing gravity anomalies. The spatial frequency is the inverse of the spatial wavelength, that is, $f_s = 1/\lambda$ in units of cycles per km. The terms wavelength and frequency are used interchangeably with an implicit inverse relationship.

Nyquist sampling theorem. The remaining question is the required sample interval. The Nyquist frequency, f_N, is the highest frequency that can be reproduced from a discrete sampling interval of a continuous function such as the gravity field. If the continuous function contains higher frequencies than the Nyquist frequency, then aliasing will occur due to inadequate sampling. Aliasing is the introduction of spurious, low-frequency anomalies into the data. For a frequency $Y > f_N$, an aliased frequency $f_N - Y$ will be introduced into the function by aliasing, or folding.

The maximum sampling interval for a time varying continuous function (time domain) containing frequencies $\leqslant f_N$ is given by the Nyquist sampling theorem:

$$f_N = 1/2\Delta T \quad \text{(time domain)} \tag{1}$$

where ΔT is the sample interval in seconds, and f_N has units of cycles per second, or Hertz. Thus, at least two samples per cycle are required. However, with gravity surveys, we work in the space domain. If the period, T, is mapped into the space domain, then $T \propto \lambda$. The Nyquist frequency in the space domain is then

$$f_N = 1/2\Delta\lambda \quad \text{(space domain)} \tag{2}$$

where $\Delta\lambda$ is the sample interval in the same length units as the spatial wavelength, λ, and f_N now has units of cycles per unit length, commonly kilometers in gravity surveys. Again, two samples per cycle is the maximum sampling interval.

If we assume that the minimum diameter of a homogeneous laccolith in our survey area is 1 km, then the associated gravity anomaly would have a frequency, f_s, of ~1 cycle per kilometer. Since the wavelength, λ, is $1/f_s$, the wavelength of the anomaly is ~1 km. For a wavelength of $\lambda \sim 1$ km, our sample interval must be $\leqslant \lambda/2$, or $\Delta\lambda \leqslant 500$ m in our example, to define a 1-km-diameter laccolith. Aliasing occurs if a larger sample interval is used. The relationship for a two-dimensional profile (i.e., signal varies in amplitude in one spatial direction only) is shown in Figure 27a and 27b.

The above approach is satisfactory if the dimension of the

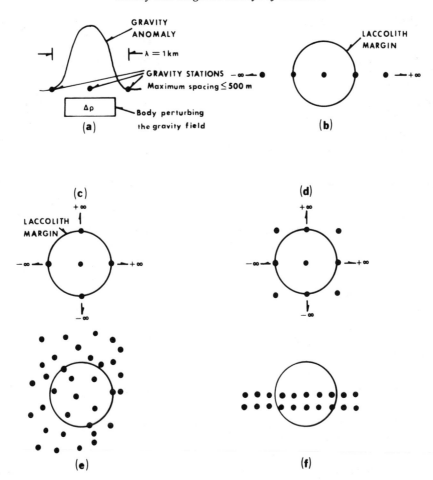

Figure 27. Illustration of the Nyquist sampling theorem in two and three dimensions in the space domain. a. Profile view of body perturbing the gravity field and resultant gravity anomaly. b. Plan view of same body. Maximum sampling interval is one-half the lateral dimension of the body for a two-dimensional profile. c. Map view. One profile perpendicular to original profile adds two points but does not satisfy Nyquist theorem. d. Map view. Two additional profiles to sample margins adds four points. Hence, nine stations are required to satisfy Nyquist theorem in three-dimensions for uniform grid of stations. e. Map view. Randomly spaced stations with average coverage of ⩾1 station per 0.1 km² over a body with ~1 km² surface area will also satisfy Nyquist theorem. f. Map view. Adequate number of stations with nonuniform spatial distribution does not satisfy Nyquist theorem.

body perpendicular to the direction of the survey profile is much greater than the dimension in the direction of the profile and the profile is far from the ends of the body. However, laccoliths are not two-dimensional bodies, and to uniquely determine the density contrast we must constrain the laccolith in three dimensions. To do this, additional data is required. An obvious step is to take another profile perpendicular to the first as shown in Figure 27c. This requires acquisition of an additional two stations. However, we have not yet sampled the other margins. To sample these margins, we must take an additional two profiles perpendicular to two additional points of our original profile as shown in Figure 27d. A total of nine stations on a uniform grid are thus required to minimally define a three-dimensional body associated with a

1 km wavelength anomaly according to the Nyquist sampling theorem.

If we state the problem in terms of areal coverage for a body covering an area of about 1 km² in map view, the spacing must be greater than one station per 0.1 km². Hence, the stations may be randomly scattered over the body, as shown in Figure 27e, but cannot be grouped in any systematic fashion (Fig. 27f) other than a uniformly spaced grid if the Nyquist theorem is to be satisfied.

Because we must also know the value of the regional, or background, field if we are to determine the anomaly (i.e., the variation from the regional field), sampling should continue to ± ∞ in all directions. Since that is not practicable, sampling is usually continued at least 5 to 10 radii away from the body.

AEROMAGNETIC SURVEYS OF LACCOLITHS

During the course of this investigation a substantial number of aeromagnetic maps of laccolithic areas were located. However, for the following reasons I am unable to use any of these surveys for the structural analysis of laccoliths.

In contrast with gravity surveys, which respond to the entire rock mass, magnetic surveys measure only a few minerals in the rock, principally magnetite. In the two laccolithic areas I am familiar with, where the magnetite relations to the laccolith have been determined, the iron-rich deposits postdate the emplacement of the laccolith. The first of these areas are the well-known magnetite deposits in the Three Springs district, Utah (Mackin, 1947b). The magnetite there is concentrated at the margins of the laccolithic intrusions by postemplacement processes. The second area is the Solitario laccolith in the Big Bend region of Trans-Pecos Texas, where Corry and others (1988) have shown that the granite forming the bulk of the laccolith has effectively zero magnetic susceptibility while the iron-rich intrusions have K-Ar ages up to 10 Ma younger than the laccolith. If these relations hold elsewhere, it is evident that magnetic surveys will only reflect postemplacement relations.

Secondly, gravity surveys measure difference in mass, a scalar. However, magnetic surveys measure intensity of magnetization, a vector. In modelling magnetic anomalies associated with intrusive bodies, the assumption is made that the observed magnetization is entirely induced by the present magnetic field of the earth. However, the magnetite in a forcible intrusion, such as a laccolith, cools through its Curie temperature after emplacement, and thereby acquires a thermal remnant magnetization in the direction of the earth's magnetic field at the time it cools. The magnetic field measured is thus the vector sum of the remnant magnetization acquired during cooling and the magnetization induced by the present earth's field. If the laccolith cooled during a period when the earth's magnetic field was reversed, the vectors may be in opposite directions, and interpretation may require careful paleomagnetic studies. Further, laccoliths are multiple-phase intrusions, and one intrusion may cool during normal polarity while the next impulse of magma might cool during a polarity reversal. To my knowledge, paleomagnetic studies have only been done on a very few of the laccolithic groups for which aeromagnetic data are available. The most recent paleomagnetic investigation of laccoliths is the work of Jackson and Pollard (1988) in the Henry Mountains, Utah.

A third problem involves aliasing due to improper survey design or effects of topography. Reid (1980) has shown that to avoid aliasing, the flight-line spacing must not be greater than twice the flight elevation for a total field survey. If individual anomalies are to be modeled, the flight-line spacing should not exceed one-half the flight elevation. In rough terrain, exemplified by laccoliths, these constraints usually cannot be met. An example of the problems that may arise is the attempted structural interpretation from an aeromagnetic survey presented by Affelck and Hunt (1980) for the Henry Mountains. This survey was flown at a constant barometric elevation of 3,660 m with a flight-line spacing of 1,600 m, which more than satisfies the conditions stated above in principle. However, the principal peaks in the area rise to 3,268 m for Mount Hillers, 3,466 m for Mount Pennell, and 3,507 m for Mount Ellen. Thus, the terrain clearance decreases to as little as 153 m over Mount Ellen. From an ideal ratio of terrain clearance versus line spacing of $\geqslant 1$, the ratio decreases to 0.1 over Mount Ellen and Mount Pennell and to 0.25 over Mount Hillers. From the work by Reid (1980, his Fig. 1), it can be determined that over Mount Ellen and Mount Pennell, up to 60 percent of the power spectrum will be aliased in the aeromagnetic survey used by Affleck and Hunt (1980). Over Mount Hillers, up to 20 percent of the power spectrum will be aliased. It is in the vicinity of these peaks that they are attempting to make detailed structural interpretations. Aliasing makes such interpretation impossible.

I have attempted the interpretation of a substantial number of similar surveys in Colorado and, in my experience, all that can be reliably located in such uncorrected aeromagnetic surveys is the mountain peaks. Bhattacharya and Chan (1977) have presented a method of correcting for such terrain effects. An example of the use of the method for the San Juan Mountains, Colorado, is presented by Wynn and Bhattacharya (1977). Without such corrections, quantitative interpretation of aeromagnetic surveys in laccolithic areas should not be attempted. Even with corrected data, the first two objections to the use of magnetic data for structural analysis of laccoliths are still valid. In short, structural analysis of laccoliths with magnetic data is not a valid use of the magnetic method except in special circumstances.

GRAVITY SURVEYS OF FELSIC LACCOLITHS

Presented below are the results of gravity surveys of 13 felsic laccolithic groups. In the Henry Mountains of Utah and the Big Bend region of Texas, the results are based principally on my data. In the Little Belt Mountains, Montana, and the Abajo Mountains, Utah, I have made reconnaissance surveys and combined my data with preexisting data principally from the U.S. Geological Survey. I made an independent gravity survey in the Sundance area of Wyoming in 1973, but it nearly adjoins on the east with the gravity and aeromagnetic maps of Kleinkopf and Redden (1975) covering the Black Hills of South Dakota. Thus, I have treated this as one area. The sources of the data used in making the gravity maps are indicated on each map. Geology for each area can be established by reference to Appendix B.

Henry Mountains, Utah

The geology of the Henry Mountains has been mapped by Hunt and others (1953), Morton (1983), Jackson (1987), and Jackson and Pollard (1988). Their work is used as a basis for the following comments. The first geophysical investigation of the Henry Mountains was done by Avakian (1970), but I have not been able to recover his data. Case and Joesting (1972) made a gravity and aeromagnetic survey of the central Colorado Plateau. Their magnetic survey extends over the Henry Mountains, but the gravity survey stops along the eastern margin. I have extended their gravity survey to the west; the combined results are shown as a complete Bouguer anomaly map in Plate I. The DMA gravity base station used for this survey is located below the benchmark on the front of the Hanksville, Utah, elementary school.

There are at least 12 anomalies shown on Plate I. However, in remote areas the data density is as low as one station for some anomalies. Only Mounts Hillers, Pennell, and Ellen are adequately sampled. The true amplitude and wavelengths of any anomalies without adequate sampling is unknown.

Starting with the southernmost major laccolith, Mount Holmes, along profile A–A'–A" on Plate I it can be seen that Mount Holmes has no anomaly with a wavelength corresponding to the laccolith. However, sampling is inadequate to completely define the Mount Holmes laccolith. The apparent high spatial frequency anomaly is possibly associated with fracturing of the roof rock, which is largely intact over Mount Holmes, or it could be due to errors in the data (e.g., terrain corrections or aliasing).

Continuing northwest across Mount Hillers along A–A', any possible anomaly is lost in the "noise" in the southeast-northwest profile. In the northeast-southwest profile, B–B', an apparent anomaly with an amplitude of about +9 mGal, can be seen. However, when the regional field is removed, the amplitude of the anomaly varies from +3 to –4 mGal. Also, the wavelength of the anomaly no longer correlates with the size of the intrusion. The regional gravity field is constrained by the requirement that it be equal at the crossing of A–A' and B–B'.

By reference to the geologic map of Hunt and others (1953) it can be seen that the major gravity high near point C on Plate I correlates with outcrops of denser Jurassic units. From point B going southwest along the profile B–B', the gravity high correlates with the surface exposure of dense Jurassic formations. Morrison Formation, with a relatively low density, is encountered northwest of Mount Hillers, which probably accounts for the observed negative anomaly. Southwest of Mount Hillers the low-density units of the Mancos Shale produce another negative anomaly. This anomaly, which B–B' cuts at the southern end, is possibly associated with coal beds and other low-density units in the Mancos Shale, which is exposed on Cave Flat where the anomaly is centered. Given the variation in density of the exposed strata and its correlation with the gravity anomalies in profile B–B', together with the lack of any anomaly along profile A–A', it is apparent that the Mount Hillers laccolith has no resolvable density contrast.

Mount Pennell is located at A' on Plate I. From the profile A–A'–A" it is obvious that no gravity anomaly with a wavelength approximating that of the intrusion is present. Hunt and others (1953) found that the diorite porphyry of the laccolith (they call it a stock) is cut by a monzonite porphyry, which may account for at least one of the high frequency anomalies observed. Deformation and fracturing of the roof rock, which is still intact

on the flanks, probably accounts for the remaining high frequency anomalies, although data errors may be included. However, it is apparent that the main bulk of the intrusion must have a density contrast near zero. The dimensions of the intrusion at Mount Pennell are such that the volume of the intrusion must be on the order of at least 50 km^3. With that volume of rock, any significant density contrast would be obvious in the gravity survey.

Continuing north from A' toward A" on Plate I, the next feature of note is the Mount Ellen stock. Gilbert (1877, p. 48) noted that ". . . there is no evidence of a great central laccolite, such as the Hillers and Pennell clusters possess. . ." The interpretation of this intrusion as a stock by Hunt and others (1953) is confirmed by the gravity data, which show a 16 mGal positive anomaly. The concept of spatial wavelengths in geophysical data is illustrated by the anomaly outlined in profile C–C' on Plate I. Note that the data density is high within the anomalous area on Mount Ellen, and that the high frequency "noise" present in both profiles A'–A" and C–C' is overshadowed by the anomaly. For comparison, the noise level at Mount Ellen is about equal to the noise level at A' above the Mount Pennell laccolith. Further, the anomaly over the Mount Ellen stock is evident on both profiles A'–A" and C–C'.

Gilbert (1877), Hunt and others (1953), as well as other investigators, have all remarked on the similarity of all the igneous rocks in the Henry Mountains, including those found in the Mount Ellen stock. Therefore, an interpretation of the Bromide Basin area as the top of a stock precludes *any* interpretation of Mount Hillers, Mount Pennell—and presumably Mount Ellsworth and Mount Holmes—as stocks, since no similar anomalies are associated with these features. The point is critical because Hunt and others (1953) and Hunt (1980) interpreted all of these features as stocks. On the basis of their interpretation, they proposed that laccoliths grow as radial, tonguelike bulges from the flanks of a central feeder stock. Other workers, notably within the U.S. Geological Survey, have expanded the hypothesis of Hunt and others (1953) and applied it in several other laccolithic groups. The available geophysical evidence contradicts their central feeder stock hypothesis, and recent mapping by Jackson and Pollard (1988) shows that the central peaks in the southern Henry Mountains are laccoliths.

Continuing north on A'–A" from the Mount Ellen stock, a large negative anomaly with an amplitude of –23 mGal is found. The anomaly correlates spatially with outcrops of low-density, coal-bearing Ferron Sandstone and other members of the Mancos shown on the geologic map of Hunt and others (1953), though deeper strucure must be present to account for the observed amplitude.

The Henry Mountains group contains many more laccoliths (Appendix B) than those discussed. A few of the larger bodies are shown on Plate I. Where gravity stations have been located on smaller laccoliths, no resolvable density contrast was found since no gravity anomalies correlate with these bodies, although as discussed above, a low signal-to-noise ratio and aliasing prevent a unique interpretation of these small bodies.

Big Bend group, Texas–Mexico

I had previously investigated the Solitario (Corry, 1972; Corry and others, 1988) in Trans-Pecos Texas. Only a portion of this laccolith is exposed, but the structural relations indicate that the Solitario is 12 km in diameter and 1.6 km thick. Thus, the total volume of the intrusion is on the order of 100 km³. The level of erosion is such that the roof rock of the laccolith has been almost entirely removed. The present surface is nearly at the same level as the pre-intrusion surface. The pre-intrusion surface outside the dome has been largely preserved under a layer of post-laccolith extrusive volcanics. Quaternary erosion has resurrected the Eocene topography, exposing the striking feature observed today.

The result of the gravity survey of the Solitario and vicinity is shown on Plate II. The DMA gravity base station for the survey is located in the Alpine, Texas, post office. From the profiles A–A′ and B–B′, it is evident that no gravity anomaly is associated with the Solitario. In the southwestern area of Plate II, steep regional gradients mark the eastern boundary of the Basin-and-Range Province.

Given the dimensions of the Solitario determined by the field work, a gravity anomaly of 6 mGal would result if the density contrast were as little as 100 kg m⁻³. A circular anomaly of this amplitude would be obvious on profiles A–A′ and B–B′. Corry and others (1988) have done an extensive study of the density of the sedimentary rocks and the laccolith in the Solitario, Trans-Pecos Texas. On the basis of several hundred density determinations, we obtained a weighted mean density for the sedimentary rocks of about 2,600 kg m⁻³, and a mean density for the laccolith, sampled vertically in a drill hole 733 m deep, of 2,574 kg m⁻³. Even though the volume of the intrusion beneath the Solitario is on the order of 100 km³, the density contrast is below the resolution of the gravity survey. That no gravity anomaly is found associated with such a large volume of intrusive igneous rock is evidence for the validity of Gilbert's hypothesis that emplacement of laccoliths is at the neutrally buoyant elevation.

No anomaly is apparent for the more deeply buried laccolith beneath the Terlingua anticline. A substantial negative anomaly is centered near Primero dome (Plate II), but the wavelength of the anomaly, and the geologic dimensions of the dome given by McKnight (1970), clearly show that the gravity anomaly is unrelated to the dome. The area around Primero dome was a major extrusive center in the Oligocene and Miocene, and the anomaly may be related to this activity.

Abajo Mountains, Utah

Case and Joesting (1972) reported a gravity anomaly of 8 to 12 mGal associated with the Abajo Mountains. At the scale of their map (1:250,000) it was difficult to establish any association with the laccoliths in the Abajo group. To solve the resolution problem, a reconnaissance gravity survey was done. The combined gravity data is shown on Plate III. However, even with the

additional data, none of the laccoliths are adequately sampled in this survey. The DMA gravity base station used for my survey is located at the National Geodetic Survey bench mark number K24 just south of Mexican Hat, Utah.

No gravity anomalies correlate with the laccoliths along profiles A–A′ and B–B′ on Plate III. A positive anomaly with an amplitude of about 5 mGal is centered between Jackson Ridge and Twin Peaks laccoliths on profile B–B′. However, the wavelength does not correlate with the dimensions of the laccoliths. Case and Joesting (1972, p. 25) state that ". . . it is very likely that a rather large igneous body is concealed at depth, or that the laccoliths were emplaced at a site which is underlain by dense Precambrian rocks." Another possibility is that the anomaly is the result of aliasing. Aliasing may mask the high frequencies because the station density is so low. The possibility of significant aliasing is rejected on the basis that no similar anomaly is found along profile A–A′, which crosses several large, closely related laccoliths. Thus, the conclusion that the laccoliths of the Abajo Mountains have a density contrast of near zero seems inescapable.

Witkind (1964a) has extended the central stock hypothesis of Hunt and others (1953) to the Abajo Mountains. The only gravity anomalies that remotely correlate with features Witkind (1964a) calls stocks are found near Shay Mountain. A single station defines a negative anomaly on the southwest side of Shay Mountain stock(?). It is probable that this anomaly is part of the negative anomaly to the south that centers over the Wilderness and Chippean Rocks. Northeast of Shay Mountain is a positive gravity anomaly of 4 to 6 mGal. However, that anomaly does not correlate in space or wavelength with the Shay Mountain intrusion, and the geophysical evidence suggests that no stock exists in the Abajo Mountains.

La Sal Mountains, Utah

Case and others (1963) conducted gravity and magnetic surveys of the La Sal Mountains. Case and Joesting (1972) supplemented their earlier survey and conclude, "The gravity anomaly of the igneous intrusions in the La Sal Mountains is obscured by anomalies related to salt thickening and to intrabasement density contrasts" (p. 24). Case and others (1963) presented an analysis of the gravity data and find a small, ≤4 mGal, positive anomaly associated with North Mountain (their profile F–F′, Plate 17). Analysis is dependent on the orientation of the profile. Perpendicular to their profile, any possible anomaly is obscured by the steep gradients associated with the Castle Valley salt anticline. The profile of the observed gravity drawn by Case and others (1963, their Fig. 42) shows no anomaly. Only by subtracting a calculated anomaly for North Mountain do they obtain their estimated gravity anomaly. The dimensions of the body are reasonably well known, and Case and Joesting (1972) have been exceptionally careful in measuring the rock densities. Also, as shown in Figure 24, the variation in density as an igneous rock cools is a function of its composition. Thus, for the diorite porphyry of the La Sal Mountains, a small positive density contrast would not violate any of my assumptions. Nevertheless, it seems unlikely that the density contrast between North Mountain and the surrounding country rock differs from zero by even as much as the densities measured by Case and Joesting (1972) would indicate, since no distinct gravity anomalies are visible.

Other positive gravity anomalies are found just east of Mount Mellenthin, between Mount Mellenthin and Mount Tukunikivatz, and to the west of La Sal Pass. The closure of these anomalies is on the order of 2 mGal at a wavelength on the order of the dimensions of the laccoliths. The larger positive anomaly, with which these higher frequency anomalies are associated, extends to the west across Black Ridge. While the wavelength of these anomalies is on the order of the dimensions of the laccoliths, the closure of the anomalies does not coincide with the positions of the laccoliths. Hence, the gravity anomalies are, at best, only indirectly related to the laccoliths.

The lack of correlation of the laccoliths with gravity anomalies indicates that the density contrast is close to zero with a possible maximum $\Delta\rho$ of +100 kg m^{-3}. One station atop Round Mountain in Case and others (1963) indicates an anomaly of 3 to 4 mGal for this small punched laccolith. However, in Case and Joesting (1972), the amplitude of the anomaly has decreased to <2 mGal. It is probable that terrain effects influence the accuracy of this station, and this anomaly(?) has been ignored.

As in the Abajo Mountains, no evidence for a deep-seated stock as described by Hunt (1958) can be found in the gravity data.

Sleeping Ute Mountain, Colorado

Sleeping Ute Mountain lies on the extreme eastern boundary of the area surveyed by Case and Joesting (1972). A gravity anomaly, with closure of as much as 8 mGal, is shown centered over Hermano Peak on their Plate 1. It is possible that the gravity anomaly indicates a stock at Hermano Peak, as proposed by Ekren and Hauser (1965). However, no gravity data exists to the south, and much of this anomaly may be an artifact of the contouring since the regional gradient to the north is steep. No gravity anomalies correlating with the laccoliths are found elsewhere on Sleeping Ute Mountain, although data density admittedly is poor. The density contrast of the laccolithic intrusions is considered to be near zero based on the scanty data of Case and Joesting (1972).

Carrizo Mountains, Arizona

Case and Joesting (1972) report a gravity high of 4 to 8 mGal. Plouff (1958, as quoted by Case and Joesting, 1972) interpreted the large gravity anomalies found in association with the Carrizo Mountains and Defiance Uplift as the expression of rock units of different densities within the underlying Precambrian basement. Thus, individual laccoliths have little, if any, density contrast with the country rock.

Little Belt Mountains, Montana

Kleinkopf and others (1972) made a gravity and aeromagnetic survey of a portion of the Little Belt Mountains. I made an additional gravity survey in the northern Little Belt Mountains. The surveys are complementary, and the combined results are shown on Plate IV. The DMA gravity base station used for my survey was located at the Stanford, Montana, airport.

The combined surveys still lack extension in the east-west direction, and this lack makes interpretation difficult. For the Pioneer Ridge dome (Plate IV), some high-frequency anomalies associated with dense Precambrian roof rocks are present. The Barker laccolith, which is well exposed, shows no deviation from any reasonable regional gradient. The Otter Creek laccolith is similarly transparent to the gravity survey. Thus, none of these features has any significant density contrast with the country rock. However, the anomaly associated with Limestone Butte dome is distinctively of the proper wavelength to correlate with the dome. The lack of east-west extension of the survey is particularly critical for an objective interpretation, since the anomaly may extend westward. Based on their magnetic data, Kleinkopf and others (1972) suggested that Limestone Butte is underlain by a laccolith 1.2 km thick with a diameter of 6.4 km. A stock 2.1 km in diameter extends to depth beneath the laccolith. My gravity data—admittedly closure is on a single station—shows a gravity high. The center of the gravity high corresponds to the center of the magnetic anomaly shown by Kleinkopf and others (1972). From drill hole data, the top of the laccolith is known to be >0.6 km beneath the present surface. This depth of burial would smooth and broaden the anomaly observed at the surface. In contrast, the gravity anomaly shown on Plate IV has a very high spatial frequency where closure is indicated. The high spatial frequency suggests a near-surface source with a diameter of ≤2 km. The remainder of the gravity anomaly at Limestone Butte, which has a wavelength of 6 to 8 km, appears to be the eastern nose of an anomaly that extends westward beyond the data limits.

Based on the present limited data, I conclude that the Limestone Butte laccolith has zero, or very low, density contrast in conformance with the observations over the other laccoliths in the group. Secondly, the gravity data suggests that the stock postulated by Kleinkopf and others (1972) cuts the laccolith and lies nearer the surface than the laccolith.

The gravity data do not indicate any continuation with depth of the Hughesville stock(?) proposed by Witkind (1973).

Black Hills, South Dakota–Wyoming

A gravity survey of the Sundance area on the western margin of the Black Hills in Wyoming was done as part of the present study. The results of this survey are shown on Plate V. The DMA gravity base station for my survey is located in the Sundance, Wyoming, post office. Kleinkopf and Redden (1975) have published gravity and aeromagnetic maps of a large portion of the Black Hills in South Dakota and south of the Sundance area in Wyoming. Thus, our maps are complementary but there is no overlap. Direct comparison between the gravity maps is not possible because Kleinkopf and Redden (1975) use a Bouguer density of 2,430 kg m^{-3} while I use the standard value of 2,670 kg m^{-3}. Also, it is probable that they calculated theoretical gravity from an older equation.

From Plate V it can be seen that all the laccoliths in the Sundance area cause a small variation in the gravity field. The cover is intact over all the laccoliths except Sundance Mountain. The other laccoliths, with intact roof rock, exhibit small (~1 mGal) negative anomalies. Only over Green Mountain does the anomaly have significant amplitude (–2 mGal). It is easily seen (A–A' on Plate V) that the wavelength of the anomaly over Green Mountain is much less than the dimensions of the laccolith, and the anomaly does not directly correlate with the intrusive body. The negative gravity anomaly does correlate with a well-exposed crestal graben.

The small negative variations associated with the other laccoliths are probably the result of fracturing the roof rock during the growth phase. As the wavelengths of these anomalies do not correlate with the intrusions, and no anomalies are apparent at a wavelength comparable to the dimensions of the laccoliths, I conclude again that the bulk density contrast of the laccoliths in the Sundance area is approximately zero.

In the much larger area of the Black Hills surveyed by Kleinkopf and Redden (1975), there is generally no correlation between the laccoliths and the gravity anomalies. They attribute a small positive (~2 mGal) anomaly at Bear Butte to the results of differential erosion between denser, more resistant Paleozoic carbonate rocks and the lighter, more easily eroded overlying sequence of shales, sandstones, and evaporites. They also point out that at Bear Butte igneous rocks exposed ". . . immediately east of the main structure appear to have no gravity expression, although this may be the result of insufficient gravity control." Other, much larger laccoliths, do not exhibit any correlation with the gravity anomalies. The density contrast must therefore be near zero.

The choice by Kleinkopf and Redden (1975) of such a low density (2,430 kg m^{-3}) for the Bouguer correction affects the appearance of many of the contours but not the conclusions.

Bearpaw Mountains, Montana

Peterson and Rambo (1967, 1972) have done a gravity survey of the Bearpaw Mountains. All of the laccoliths in the Bearpaw Mountains shown in Appendix B were plotted on a copy of their map. No correlation between any of the anomalies and the laccoliths is evident. Data density is generally inadequate to define the dimensions of individual laccoliths so any high frequency anomalies possibly associated with the laccoliths are aliased. However, it is clear that the larger laccoliths in this group cannot have any significant density contrast.

San Francisco Mountains, Arizona

J. D. Hendricks (personal communication, 1975) made a gravity survey of the San Francisco Mountains. Two of the three laccoliths in the San Francisco Mountains are covered by Hendricks' survey. Marble Mountain causes no deviation in the gravity contours drawn by Hendricks. Several stations are located on Marble Mountain, and there is little doubt that the density contrast is near zero. Only the northern portion of Elden Mountain is covered by the gravity survey. A small (~2 mGal) positive anomaly appears to be associated with Elden Mountain. This is not surprising since Elden Mountain apparently broke through to the surface during the growth cycle (Robinson, 1913). A small positive anomaly is expected if the roof does not remain intact during growth.

San Juan Mountains, Colorado

Plouff and Pakiser (1972) published a gravity survey of the San Juan Mountains, Colorado. They find (p. B188) that "The widespread intrusive rocks in the west part of the area do not generally correlate with gravity highs. These rocks, for the most part, are associated with laccoliths rather than with volcano centers that typify the eastern gravity highs and sills." No correlation of the anomalies, of course, means no density contrast.

Death Valley, California

Hunt and Mabey (1966) have included a gravity survey in the geologic map (their Plate 3) of a portion of the Death Valley region in California. Two large laccoliths, Hanaupah Canyon and Skidoo, are covered by their survey. Neither of these intrusive bodies shows any significant perturbation of the gravity field. Hence, the density contrast is close to zero.

Crazy Mountains and Little Rocky Mountains, Montana

Bonini and others (1967) have done reconnaissance gravity surveys of the Crazy Mountains and Little Rocky Mountains of Montana. In the Crazy Mountains, both the Big Timber and Loco Mountain stocks have associated gravity highs, although Bonini and others (1967) find that the idea of separate stocks is inconsistent with the gravity data. The laccoliths have no apparent gravity anomalies, although this may be in part due to low data density.

The gravity data in the Little Rocky Mountains is limited to a single northeast to southwest profile, which shows an unusual negative anomaly associated with the Little Rocky Mountains. However, from the profile and models drawn by Bonini and others (1967) it can be seen that the wavelength of the anomaly greatly exceeds the dimensions of any of the laccoliths described in the Little Rocky Mountains.

GRAVITY SURVEYS OF MAFIC LACCOLITHS

In contrast with the zero density contrast associated with felsic laccoliths, mafic laccoliths exhibit a positive density contrast with respect to the country rock. From the data in Figure 24, and the associated discussion, a positive density contrast is expected with mafic laccoliths. Available gravity surveys of mafic laccoliths are summarized below.

Exeter pluton, New Hampshire

Bothner (1974) made a gravity study of the Exeter pluton in southeastern New Hampshire. The pluton is 32 km long by about 7 km wide and has the approximate form of a laccolith, both from the gravity study and structural relationships. The composition of the pluton is transitional between the mafic and ultramafic intrusives reported on below and the felsic laccoliths described above. The principal rock type is diorite, with a complete variation from gabbro to quartz monzonite existing within the pluton. The magma intruded Silurian(?) metasedimentary rocks with a density of 2,670 to 2,700 kg m^{-3} (Bothner, 1974). The diorite presently has a mean density of 2,800 to 2,850 kg m^{-3} with a range from 2,600 to 3,100 kg m^{-3} (Bothner, 1974), reflecting the compositional variation in the pluton.

A positive gravity anomaly with a maximum amplitude of +15 mGal is associated with the pluton (Bothner, 1974). He uses a density contrast of +150 kg m^{-3} to model the pluton, with a resultant thickness from his models of 1 to 3 km.

Mustang Hill laccolith, Texas

Greenwood and Lynch (1959) made a gravity survey over a small alkaline basalt laccolith in Uvalde County, Texas. A positive 2.2 mGal gravity anomaly is associated with the intrusion. From density measurements of the sedimentary rocks and the intrusive rock, Greenwood and Lynch (1959) estimate a density contrast of +500 kg m^{-3}. This density contrast yields a maximum thickness of 105 m for the laccolith based on a Bouguer slab approximation. The predicted density contrast for an epizonal mafic laccolith emplaced at its neutrally buoyant elevation would be approximately +100 kg m^{-3} after solidifying. A density contrast of +100 kg m^{-3} would require a laccolith 250 to 500 m thick. Based on experience in west Texas in similar stratigraphy, I suspect Greenwood and Lynch's (1959) measurements underestimate the density of the sedimentary rocks and that the density contrast is ≤200 kg m^{-3}. However, the laccolith is completely unroofed, and no independent estimate of the thickness from the structural relief appears possible.

The structure sections shown by Greenwood and Lynch (1959) require the superior beds to turn up around the laccolith and the inferior beds to dip down below the laccolith. Such behavior is not characteristic of epizonal laccoliths.

Even with the extreme density contrast of 500 kg m^{-3} they used, the gravity survey indicates a laccolith considerably wider, and twice as thick as the structure section. If the laccolith is a Christmas-tree laccolith, both the geology and geophysics can be reconciled, and a lesser density contrast would be required.

Trompsburg lopolith, South Africa

Buchmann (1960) made a gravity and magnetic study of a large lopolith in the vicinity of Trompsburg, Orange Free State, South Africa. The intrusion is known from drilling to be gabbro, olivine gabbro, anorthosite, and norite. The mafic rocks of the intrusion are assumed to have an average density of 3,000 kg m^{-3}, and the country rock is assumed to be granite with a density of 2,650 kg m^{-3}. The density contrast would then be +350 kg m^{-3}. The gravity anomaly has a peak value of +99.5 mGals.

Mara laccolith, Tanzania

Darracott (1974) made a gravity study of the Mara laccolith in northern Tanzania. The intrusion is assumed to be gabbro. A residual gravity anomaly of +25 mGals is associated with the body. The country rock is granite with an assumed density of 2,620 kg m^{-3}, and the density of the intrusion is assumed to be 3,000 (+120/−150) kg m^{-3}. The density contrast would then fall in the range +230 to +500 kg m^{-3}. The range of thicknesses for these density contrasts would be 1,200 m to 2,600 m using a Bouguer slab approximation.

Channel Islands, United Kingdom

Briden and others (1982) have done a gravity and magnetic study of the Channel Islands in the United Kingdom. Positive gravity anomalies over the northeast portion of Guernsey Island and the southeast portion of Jersey Island are attributed to laccolithic intrusions. The St. Peter Port gabbro at Guernsey is assumed to have a density contrast of +270 kg m^{-3}, and the mafic rocks at Jersey are modeled using a density contrast of +200 kg m^{-3}.

Joutsenmaki intrusive, Finland

Parkkinen (1975) reports on the Joutsenmaki intrusive in the Haukivesi area of Finland. The body has the shape of a sill or laccolith 18 km long by about 6 km wide. Parkkinen (1975, p. 37) includes a portion of a Bouguer gravity anomaly map showing an anomaly of approximately +35 mGals centered over the northwestern portion of the Joutsenmaki intrusive. No interpretation of the gravity data is included in Parkkinen's (1975) paper, and it is possible the gravity anomaly is unrelated to the Joutsenmaki intrusive. However, should the gravity anomaly be a result of the intrusion, a density contrast of +200 to +400 kg m^{-3} would be geologically realistic.

SUMMARY OF GRAVITY DATA

The gravity surveys demonstrate that where felsic laccoliths have intruded simple stratigraphic sequences in the shallow crust, and regional metamorphism has not obscured the emplacement process, no significant density contrast between felsic laccoliths and the country rock presently exists. The sample of felsic laccoliths that have gravity surveys showing essentially zero density contrast is fairly large. The reasonable conclusion is that the dominant factor controlling the level of emplacement is the density contrast, as Gilbert (1877) originally proposed a century ago.

I have suggested that felsic laccoliths emplaced at sufficient depth may have a negative density contrast after cooling. Negative gravity anomalies reported by Lind (1967) and Laurén (1970) over laccolithic bodies with the dimensions of batholiths (i.e., tens of kilometers across and 5 to 15 km thick), emplaced in metamorphic country rock are thus consistent with Gilbert's hypothesis.

While laccoliths formed from basic magmas do have a positive density contrast, as expected, a problem is apparent with the magnitude of the density contrast found. From the work of Murase and McBirney (1973), illustrated in Figure 24, basic magmas are expected to increase in density 150 to 200 kg m^{-3} during solidification. Shallow, mafic intrusions emplaced at their neutrally buoyant elevation should thus have a density contrast of approximately +100 kg m^{-3} at present. However, the gravity surveys of mafic laccoliths cited above assume density contrasts of 150 to 500 kg m^{-3}. I assume that +500 kg m^{-3} for the density contrast is probably unrealistically high due to an underestimate of the thickness of the laccolith and the mean density of the sedimentary rocks, but even a range in $\Delta\rho$ of 150 to 400 kg m^{-3} exceeds the predicted values. The gravity surveys of mafic intrusions cited above also suggest that there is a progressive increase in density contrast as the type of magma that forms the intrusions becomes more basic.

The vapor barrier proposed by Mudge (1968), the stress barrier proposed by Gretener (1969), or Weertman's (1973, 1980) models would act to stop the ascent of the magma *below* the neutrally buoyant elevation. The present density contrast for a basic magma should then be zero or slightly negative if they increase in density by the amounts indicated by Murase and McBirney (1973). Thus, if a vapor barrier, stress barrier, or any other mechanism is stopping the ascent of the magma below the neutrally buoyant elevation, an even greater increase in density during solidification is required to account for the presently observed density contrast. It is apparent that, if the magma is not stopping its ascent at the neutrally buoyant elevation, it must be going *above* the neutrally buoyant elevation to achieve the present density contrast. Otherwise the data of Murase and McBirney (1973) regarding density variation during cooling cannot be correct.

A number of factors could be responsible for the apparent discrepancy in the present density contrast found with mafic laccoliths:

1. The assumed density contrast is too high due to an underestimate of the dimensions, notably the thickness, of these bodies.

2. The magma rose above its neutrally buoyant elevation due to tectonic overpressure or some other mechanism. This problem is reviewed by Williams and McBirney (1979, p. 60–62).

3. The density of basic magmas increases more on cooling than indicated by the limited experimental work. Basic magmas appear to have densities around 2,600 kg m^{-3} (Fig. 24), while most mafic rocks, if not vesicular, have densities around 2,900 to 3,100 kg m^{-3}. If these values represent the actual increase in density of mafic rocks during cooling, the present density contrast for mafic laccoliths should be on the order of +200 to +300 kg m^{-3} if emplaced at their neutrally buoyant elevation.

4. Basic magmas contain fewer phenocrysts on average. Hence, they are not as commonly a crystalline mush at the time of intrusion as silicic magmas. Therefore, the solidification process affects a larger percentage of the total volume of the intrusion in basic magmas. The net result would be a larger percent increase in density of mafic intrusions as they solidify.

5. The bodies described in the included investigations are not laccoliths but some other type of intrusive body. For example, Maguire (1980), in a gravity study in the fall zone in northeastern Delaware and northeastern Maryland, found the most plausible explanation for the presence of several small gabbroic bodies, with the approximate form of laccoliths, was as a result of thrust faulting of pieces of oceanic crust.

In my experience, it is probable that hypothesis 1, above, is true to some extent. In any gravity modelling there is a tendency to make the body as thin as possible. For hypothesis 2, it is difficult to see why only mafic laccoliths would be forced above their neutrally buoyant elevation. Tectonic stresses, or similar factors, should be acting on silicic magmas as well; yet they do not show any density contrast. Given the limited amount of experimental work, and the requirement for an even larger density increase if the magma is stopped below its neutrally buoyant elevation, hypotheses 3 and 4 seem the most probable explanations for the large observed density contrast between mafic laccoliths and the surrounding country rock. As pointed out in 5, some intrusions investigated may not be laccoliths at all.

While the gravity data indicates that some laccolithic areas have stocks within the group, the relation does not appear to be common. There is no geophysical evidence to indicate a cause-and-effect relationship between stocks and laccoliths, as originally proposed by Hunt and others (1953). In the Henry Mountains, the gravity data show that only the southern part of Mount Ellen (Plate I) can possibly have any continuation with depth and that the other mountains are large laccoliths, as Gilbert (1877) and Jackson and Pollard (1988) described. Also, vertical feeder dikes beneath the floors of several laccoliths have been found (e.g., see Fig. 22). As discussed above, laccoliths spread to their full diameter as thin sills. Therefore, the hypothesis of Hunt and others (1953) and Hunt (1980) that laccoliths are formed as radial tongues fed laterally from a central stock is not supported by any available data.

CHAPTER IV

THE EMPLACEMENT OF LACCOLITHS

The emplacement and growth cycles of laccoliths can be conveniently divided into four sequential stages. These stages are linked by a time progression, and the final conditions of one stage are used as initial, or boundary conditions for the next stage. The first two stages occur during the emplacement cycle, and the last two stages represent the growth cycle.

STAGE ONE. THE MOVEMENT OF THE MAGMA VERTICALLY THROUGH THE LITHOSPHERE

The generation of magma by partial melting in the mantle has recently been reviewed by Waff (1986) and subsequent articles in the same issue, but as did Gilbert (1877), I consider the question of magma generation to be beyond the scope of this report.

Stage one requires a mechanism whereby the magma can rise from its source through the lithosphere to the level of the future intrusion. Several authors have reviewed the problems associated with magma transport through the lithosphere. Shaw (1980) looked at fracture mechanisms associated with magma transport. Marsh (1982) has examined the mechanics of diapirism and stoping. Turcotte (1982) points out that the magma must migrate to the surface because of the buoyant forces associated with the magma's lower density. Present evidence from gravity surveys of felsic laccoliths and assumptions one and two (in Chapter II) require a mechanism for magma transport dependent solely on the density contrast between the magma and the country rock above the magma source. Models of magma transport based on dislocation theory derived by Weertman (1971a, 1971b) are thus particularly attractive and have the advantage of allowing calculation of the magma pressure within the dislocation crack.

Dislocation model

In Weertman's analogy, the magma acts as the extra half-plane of a dislocation and forms a crack in the country rock that can be considered to be a collection of infinitesimal edge disloca-

tions, as in Figure 28. The vertical climb force, F_z, per unit crack length is given by Weertman (1971a) as

$$F_z = (\bar{\rho} - \rho')gV. \tag{3}$$

(Note: the notation for all equations and models is given in Appendix C). The critical vertical crack length, L_c, is (Weertman, 1971a, 1971b)

$$L_c = [2VG/\pi\alpha g(\bar{\rho} - \rho')]^{1/3}. \tag{4}$$

The critical crack length, L_c, is the length at which $D(z)$ becomes negative, a physical impossibility. Hence, $2L_c$ determines the total crack length. The profile of a stationary liquid-filled crack is given by Weertman (1971b) as

$$D(z) = [(\bar{\rho} - \rho')g\alpha/G] \cdot (L_c^3/L^2) \cdot (1 + L^2z/L_c^3) \cdot (L^2 - z^2)^{1/2}. \tag{5}$$

where $D(z)$ is the lateral displacement of the crack face at z for $-L < z < L$, with the condition

$$D(z) = 0 \text{ for } |z| > L_c.$$

The parameter V is a measure of the volume of the crack per unit length and has units of length cubed per unit length. The total crack length, $2L_c$, depends directly on V, which can be calculated from the relations given by Weertman (1971b). Weertman (1971b, p. 8551) presents a table of values for L_c and $\bar{D}$, where $\bar{D}$ is the average crack or dike width for V ranging from 10^{-1} to 10^3 m^3 m^{-1}; part of this table is duplicated in Figure 28. $\bar{D}$ ranges up to 4.4 m for a magma viscosity of 100 Pa s and density contrast of 400 kg m^{-3} chosen by Weertman (1971b) to obtain V for "lava rock." For mafic intrusions these are probably reasonable values for $\bar{D}$ and are within the range given by Fujii and Uyeda (1974) for mafic dikes. With the notable exception of pegmatitic dikes, most felsic dikes have a width $\leqslant 5$ m. The width of silicic feeder dikes for laccoliths that I have observed ranges

$V(m^3 m^{-1})$	$L_c(km)$	$\overline{D}\ (m)$
10^3	11	4.4
10^2	5.3	0.95
10^1	2.4	0.20
10^0	1.1	0.044
10^{-1}	0.52	0.0095

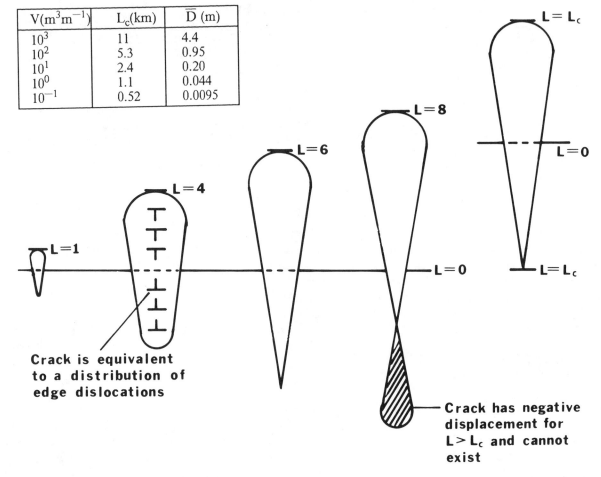

Figure 28. Profiles of liquid-filled cracks of increasing half-lengths, L, in the interior of a plate, after Weertman (1971a). Profiles can be calculated using equation 5. Included table from Weertman (1971b) lists critical crack length, L_c, and average crack displacement, $\overline{D}$, for several values of V for "lava rock" filled cracks.

from 1 to 4 m. Dikes of the same rock type as the laccoliths and exposed within laccolith groups have a similar thickness range.

A basic assumption of Weertman (1971a) is elastic, brittle behavior. Experiments by Pollard (1973) indicate that dike widths in ductile material should be significantly greater than in brittle material. Therefore, Weertman's model should only be strictly applied to dikes that penetrate brittle material. However, feeder dikes, as shown in Figure 22, may show plastic deformation on the margins without significant thickening of the feeder dike.

Magma is generally assumed to originate at depths at least on the order of the thickness of the crust in continental areas (30 to 50 km, see review by Waff, 1986). It is shown below that the magma driving pressure in a gravity-driven system is a function of the total crack length, $2L_c$, and that the magma driving pressure, P_d, must be at least one-half the lithostatic pressure, P_l, in order for a laccolith to form. Therefore, if the magma is generated at depths less than ~10 km, it would probably not form a

laccolith because of inadequate magma pressure to cause roof failure. Turcotte (1982) points out that magmas associated with island arcs must migrate upward on the order of 150 km. McBirney (1984, p. 26) argues for depths of 100 to 200 km for magma generation. An estimate of the maximum possible value for the conduit length of approximately 40 km seems reasonable. A crack length of 40 km would correspond to L_c ~20 km, $\overline{D}$ ~20 m, and V ~10^4 m^3 m^{-1} in Weertman's dislocation model. The maximum estimated value for $\overline{D}$ is five times the observed dike widths for silicic magmas, and dike widths for basic magmas are even smaller. Given depths of magma formation that are probably $\geqslant$50 km, and crack lengths, $2L_c$, of <40 km, it seems improbable that the magma conduit remains connected to the magma source all the way to the surface.

The dislocation model provides constraints on dimensions of the feeder dike in two dimensions, width and depth. It is also possible to constrain the third dimension. It is a common observation that laccoliths have a minimum diameter of approximately

1 km. Since many laccoliths are scarcely larger than this, and most feeder dikes do not extend beyond the boundaries of the laccolith, the lengths of many feeder dikes are ≤1 km. A length of ≤1 km is also in general accord with segment lengths of surface exposures of dikes in laccolithic areas. Constraining the length of the feeder dike imposes some difficulty with Weertman's theory for he assumed the crack was infinitely long (Weertman, 1971a). However, it can be assumed that if the boundaries are more distant than five times the dimension of interest they can be treated as infinitely far away (Saint Venant's principle). In the present problem the known dimension of $\bar{D}$ is ≤4 m. Thus, if the ends of the crack are 20 m distant, Weertman's theory should be valid.

Volumetric calculations for feeder dikes. Secor and Pollard (1975), Pollard and Muller (1976), Pollard (1976), and Weertman (1980) have examined the critical volume relations for a fluid-filled crack to propagate toward the earth's surface under the assumption that the crack is a dislocation. If the fluid density is less than the surrounding rock, there is a strong tendency for the crack to move upward with relatively small volumes of contained fluids.

The problem can also be examined on the basis of available field data. Since the dimensions of many feeder dikes are known within reasonable limits, the volume is readily calculated. For a feeder dike 40 km deep, 20 m wide, and 1 km long, the maximum possible contained volume is about 0.8 km^3. From the observed value of $\bar{D}$ ≤4 m, a more typical contained volume of magma in a feeder dike would be ≤0.1 km^3. These volumes are reasonably consistent with the volumes of "magmons" calculated by Scott and Stevenson (1986, p. 9290). Since the total volume of the target laccoliths for the gravity surveys is commonly >10 km^3, the total contents of a great many feeder dikes must be contained in the present laccolith.

Pseudo-Archimedian buoyancy. The force exerted upward by the crack is a result of the lower density of the magma filling the crack. Weertman's (1971a, p. 1177) succinct explanation describes the process adequately:

The force of $(\rho-\rho')$ gV experienced by the crack is identical to the Archimedian buoyancy that would act on the crack if it were a solid of density ρ' embedded in a liquid of density ρ. There is a fundamental difference, however, between these two forces. For the true Archimedian force the body force ρ' gV differs from the net force ρgV exerted on the embedded solid by the hydrostatic pressure. In the case of the liquid filled crack, the body force ρ'gV exerted on the liquid within the crack must be, and is, exactly balanced by hydrostatic forces exerted against the crack walls. *There is no net force exerted on the liquid within the crack.* The Peach-Koehler climb force is produced by a gradient in elastic strain energy within the solid plate. The origin of the true and the pseudo-Archimedian forces is the same, namely the gravity field.

Magma driving pressure. The driving force for the dislocation is given by the unmodified Peach-Koehler equation (Weertman and Weertman, 1964). For any measured value of $\bar{D}$, the total magma driving pressure can be obtained by integrating

over the total crack length given in Figure 28. Alternatively, the magma driving pressure can be estimated at any point within the crack by the following relation:

$$P_d = P_t - P_1 = [\bar{\rho}_1 \, g(2L_c + z_1) - \rho'gh] - \bar{\rho}_z gz(h) \qquad (6)$$

for Newtonian magmas, and

$$P_d = [\bar{\rho}_1 g(2L_c + z_1) - (\rho'g + 2k'/\bar{D})h] - \bar{\rho}_z gz(h) \qquad (7)$$

for a Bingham magma with a yield strength k'; where h is the height of the observer above the bottom of the crack, z_1 is the depth of intrusion or to the top of the crack, $\bar{\rho}_z$ and z(h) are evaluated at $2L_c + z_1 - h$, and L_c is determined from the relation between $\bar{D}$ and L_c as shown in Figure 28. For a magma density, ρ', of 2,500 kg m^{-3} the magma driving pressure is calculated for various crack lengths in Figure 29. In the model shown in Figure 29, $\rho' = \bar{\rho}$ at a depth of 3 km. The maximum magma driving pressure, $P_d = P_t - P_1$, occurs at the level of intrusion. Johnson and Pollard (1973) have presented similar models for estimating the magma driving pressure.

A reasonable estimate of the magma driving pressure can therefore be made from Weertman's dislocation model by measuring the width, $\bar{D}$, of feeder dikes, determining L_c as in Figure 28, and solving equation 6 or 7, as illustrated in Figure 29.

Magma incompressibility. In estimating magma driving pressure from Figure 29, it should also be noted that the magma is assumed to be an incompressible fluid, and that the magma density does not vary with depth. Any increase in density due to increased pressure will be opposed by a decrease in density due to increasing temperature with depth. The lack of superheat in the magma, evidenced by the lack of thermal metamorphism at the laccolith contact, implies that the magma must ascend adiabatically. McBirney (1984) points out that in such circumstances the decrease in density due to decreasing pressure will be close (within a factor of two) to the increase in density due to decreasing temperature. Thus, an assumption of incompressibility is a reasonable first approximation. The density of existing crystals within the magma will be relatively unaffected by changes in pressure and temperature. To whatever extent the magma is a crystalline mush, estimated to be ≥25 percent for felsic laccoliths, it will act as an incompressible fluid. Delaney and Pollard (1981) point out also that an assumption of incompressibility is justified if the change in density due to vesiculation and crystallization is negligible in comparison with the flow rate during ascent. Further, the models for laccolith formation do not appear to be particularly sensitive to the magma driving pressure, P_d, so long as $2P_d > P_1$, as derived subsequently from punched model three.

Physical implications of dislocation model. While the use of Weertman's dislocation theory is an approximation of the physical problem, and the quantities required for solution of the problem are imprecisely known, the physical implications of the model are clear:

1. More fluid magmas, usually more basic magmas, will

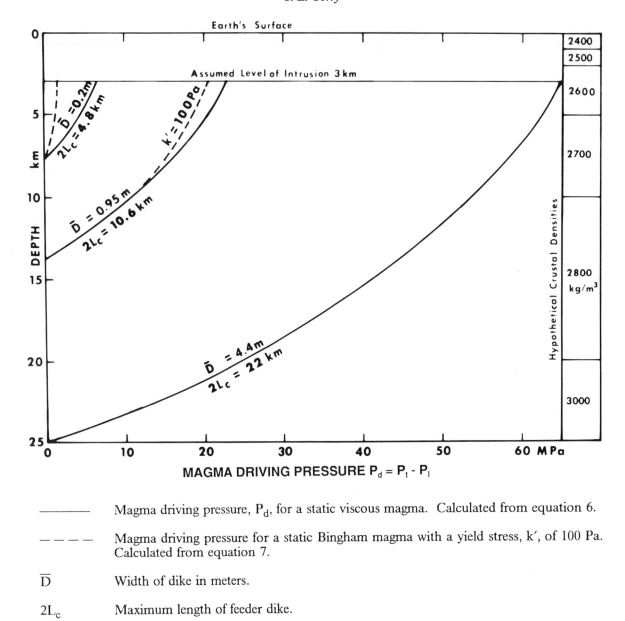

——————— Magma driving pressure, P_d, for a static viscous magma. Calculated from equation 6.

– – – – Magma driving pressure for a static Bingham magma with a yield stress, k', of 100 Pa. Calculated from equation 7.

$\overline{D}$ Width of dike in meters.

$2L_c$ Maximum length of feeder dike.

Figure 29. Magma driving pressure for a magma of density $\rho' = 2,500$ kg m^{-3} for crack lengths and widths shown. Intrusion of protolaccolith occurs at neutrally buoyant elevation, 3 km for crustal densities chosen.

have a shorter critical crack length, L_c. Hence, they will have a lower magma driving pressure, P_d.

2. Since $\overline{D}$ is directly related to L_c, a more fluid magma will have a smaller $\overline{D}$. Therefore, the effect of a finite magma shear strength, k', will be to further lower the magma driving pressure, as shown in Figure 29 and expressed in equation 7.

3. The dislocation model can be used to put approximate bounds on related magma properties, since $\overline{D}$ can be obtained by measuring the width of feeder dikes.

STAGE TWO. REORIENTATION OF THE MAGMA FROM VERTICAL CLIMB TO HORIZONTAL SPREADING

The magma, which will eventually form a laccolith, must stop at some elevation, and the flow direction must reorient from vertical climb to become a horizontal tabular sill, or protolaccolith. As Anderson (1951) pointed out, in an elastic, isotropic medium, the crack tip is so easily propagated that the dike should continue vertically until the weight of the magma above the neutrally buoyant elevation balances the magma driving pressure. Dikes that extend above the laccolith can be found (Fig. 1, bottom). Thus, some mechanism must be operating to reorient the magma at its neutrally buoyant elevation.

A laccolith emplaced significantly above the neutrally buoyant elevation must have a distinct positive density contrast with respect to the country rock and a corresponding positive gravity anomaly. Similarly, factors that might cause a felsic laccolith to be emplaced below the neutrally buoyant elevation, such as a stress barrier (Gretener, 1969), a freely slipping joint (Weertman, 1980), or possibly a high value of viscosity or magma strength, k′ (and see equation 7), would cause a felsic laccolith to have a negative density contrast with a corresponding negative gravity anomaly. Since negative gravity anomalies have not been observed with shallow felsic laccoliths, I conclude that the level of intrusion is determined predominantly by the density contrast between the rising magma and the country rock, as Gilbert (1877) originally proposed. In Weertman's (1971a, 1971b) dislocation model, the climb force (F_z in equation 3) goes to zero when $\bar{\rho} = \rho'$, and the magma will then reorient to horizontal spreading.

The neutrally buoyant elevation may be a narrow vertical range if the variation of density with depth, $d\rho/dz$, of the country rock is large. Alternatively, if $d\rho/dz$ is small, the neutrally buoyant elevation will extend over a substantial vertical range. For the crustal model given in Figure 25, the vertical range of possible intrusion is about 1 km, and over this distance the assumed weighted mean density varies <20 kg m^{-3}. Most laccoliths within laccolithic groups I have investigated appear to be emplaced in an elevation range of about 1 km. Magmas emplaced in the same area with significantly different densities (>10 kg m^{-3}) will emplace at different levels. However, a substantial variation in density implies a variation in composition. Field evidence indicates that compositional variations are often small within an individual laccolith. Commonly, the igneous rocks of an entire laccolith group will have very similar composition and, hence, similar magma density.

Previous authors have emphasized the role of the country rock ductility as a controlling factor for the intrusion. Cracks do not propagate as readily through a ductile medium as through a brittle medium (see discussion and experimental results in Pollard, 1973). However, if the medium is ductile *and* isotropic then propagation in *any* direction is impeded. Field evidence demonstrates that the sills are usually contained within the ductile beds and are not formed preferentially at either the top or base of the ductile beds. Sills migrate through beds that must have been brittle at the time of intrusion (e.g., rhyolite flows, quartz cemented sandstones, quartzite). Thus, country rock ductility does not seem to be the factor controlling reorientation or spreading of the magma.

The anisotropic nature of most ductile beds, such as shale, does not seem to be accounted for in previous models. Shales are quite strong (see data in Handin, 1966) normal to the bedding plane but possess very weak parting surfaces parallel to the bedding. Brittle beds also commonly possess weak surfaces along bedding planes, or between the flows in the case of lava beds. The presence of these strongly anisotropic layers may locally favor a given level of emplacement. However, horizontal parting surfaces, or a freely slipping joint as proposed by Weertman (1980), are so common within a given geologic section that it would be remarkable not to find one within a vertical range of, say, 100 m. A favorable horizon for sill emplacement should therefore exist at every level within the range of neutral buoyancy. Sill formation may locally be favored by the presence of strongly anisotropic beds, but if the magma is below its neutrally buoyant elevation, the magma does not reorient to a sill, except temporarily as illustrated in Figure 26.

Once the magma has reoriented from a vertical dike at its neutrally buoyant elevation, it spreads radially as a sill. Attempts to deal quantitatively with the intrusion of sills began with Anderson (1938, 1951). The stress distribution around "penny-shaped" cracks was investigated by Sneddon (1946). The growth of pressurized "penny-shaped" cracks was developed in Sneddon and Lowengrub (1969). These models were reviewed by Pollard (1973), who in turn presented a model for two-dimensional elliptical sheet intrusions. Pollard (1976) and Pollard and Holzhausen (1979) continued the investigation of hydraulic fractures in the earth's crust. Shaw (1980) reviews magma transport in fractures. As Pollard and Holzhausen (1979) point out, fractures associated with intrusions are seldom planar. This is certainly true of laccolith floors. Floors of laccoliths are corrugated; they cut up dip (Pratt and Jones, 1965) down dip (Witkind, 1964a) and exhibit other irregularities. After reorienting, the magma appears to advance in the plane of neutral buoyancy without regard to local structure. This is particularly obvious in laccoliths in the Black Hills of South Dakota, which cut laterally across dipping beds in a plane parallel to the earth's surface. Other examples of crosscutting abound.

Bradley (1965) has observed that a sill whose level of intrusion is controlled by density contrast will respond to variations in surface topography by stepping up, or down, in a series of steep dikes (see Fig. 26). This effect will be more noticeable in shallow intrusions and over depth ranges where $d\rho/dz$ is large. Since small laccoliths are more common, this effect is probably not very noticeable. Variations in the level of the floor observed in small laccoliths is probably the result of interference due to nearby intrusions (e.g., see Fig. 9), or as a result of crosscutting previously tilted beds, as appears to have occurred in the Black Hills, South Dakota, and at Bee Mountain, Texas.

Fingered sheet intrusions

After hearing the results of Dockstader's (1974) work on sill fingers associated with Shonkin Sag laccolith, Montana, I visited the Abajo and Henry Mountains, Utah, and also the Big Bend group in west Texas, where fingers are prominently visible at Bee

Figure 30. Fingers at the edge of Bee Mountain laccolith (Ti), north of Study Butte, Texas, on Texas 118. The view is looking northwest on the eastern margin of the laccolith adjacent to the highway. The country rock (Kbo, Boquillas Formation) is the light-colored rock at the base of the picture. The grooves left in the country rock are visible between the fingers, and remnants of the country rock remain trapped between fingers (Fig. 32).

Figure 31. Finger at the exposed floor of Trachyte Mesa laccolith in the Henry Mountains, Utah. View is looking east on the northwest flank of the laccolith. Hunt and others (1953) claim this intrusion was intruded laterally from Mount Hillers in part because the long axis of the laccolith is aligned in the direction toward Mount Hillers. However, the direction of flow here was perpendicular to the long axis of the intrusion, or at a right angle to their proposed direction of intrusion. In my model, I propose that the long axis represents the orientation of a post-Mount Hillers dike, which acted as a feeder for the Trachyte Mesa laccolith. Hence, flow perpendicular to the axis of the feeder dike would be expected.

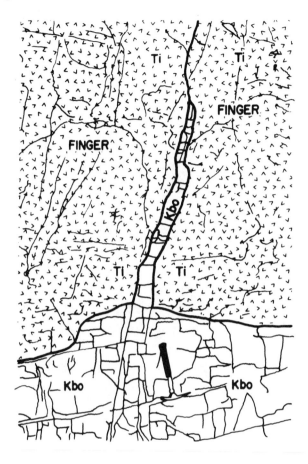

Figure 32. Remnant of country rock (Kbo, Boquillas Formation) between advancing fingers at eastern margin of Bee Mountain laccolith (Ti) north of Study Butte, Texas, on Texas 118. Beds dip southwest, and the laccolith cuts across bedding here. View is looking west on the eastern margin of the laccolith. Another view of these fingers is shown in Figure 30.

Mountain (Fig. 30). To me it appeared that the magma advanced by extending what I termed "pseudopods" after the organ of locomotion of protozoans. Dockstader (1977, 1978) and Pollard and others (1975) have presented their research on "fingered" sheet intrusions. My field observations support their conclusions.

Pollard and others (1975, p. 357) attribute the generation of fingers to instabilities on the interface between the magma and country rock. These instabilities are caused by the difference in viscosity between the magma and the country rock. Once a finger begins to grow, it extends rapidly to some limiting length. The limiting length of the finger is apparently dependent on magma viscosity (Pollard and others, 1975, p. 357–358). They observed finger lengths of 1 m in the Henry Mountains (Fig. 31), and 100 m at Shonkin Sag, which they account for in their model if the magma viscosities were in the range 1.6×10^2 to 2.5×10^7 Pa s for the Henry Mountains and 15 to 470 Pa s for Shonkin Sag. After reaching the limiting length, growth of the finger changes to a coalescence with adjacent fingers at some distance behind the finger tips, and eventually the sill face advances by the length of the finger. Thus, the sill advance resembles the locomotion of a protozoan, which might be pictured as growing radially.

The coalescence of the advancing fingers deform and brecciates the country rock between the fingers as illustrated in Figure 32 and by Pollard and others (1975, their Fig. 8). The brecciation of the country rock between fingers could partially account for the common presence of breccia, or shattered zones, associated with the roof contacts of laccoliths.

I have observed corrugations in the floors of laccoliths in the Abajo, La Sal, and Henry Mountains in Utah, and in the Big Bend group in Texas. These corrugations correspond to the path of magma fingers. They are approximately sinusoidal in cross section and tend to have a uniform wavelength as shown in Figure 33. This observation supports the prediction of Pollard and others (1975, p. 357) that a well-defined dominant wavelength of the finger spacing might develop in the intrusion. Similar corrugations can occasionally also be found in the roofs of laccoliths. Johnson and Pollard (1973) describe such grooves in the roof of Trachyte Mesa in the Henry Mountains, Utah.

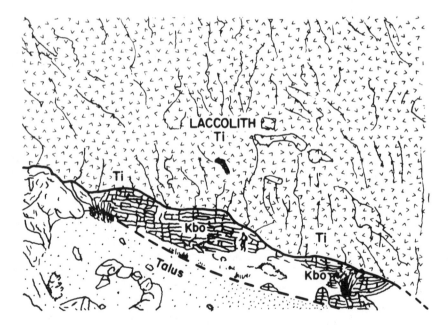

Figure 33. The floor of Indian Head laccolith is exposed at Joe Black spring near Study Butte, Texas. The view is looking north. The finger spacing is uniform, and a dominant wavelength for finger spacing was developed. The laccolith is floored in Cretaceous Boquillas Formation (Kbo). See figure for scale.

CHAPTER V

THE GROWTH OF LACCOLITHS

STAGE THREE. CESSATION OF HORIZONTAL SPREADING AND COMMENCEMENT OF THICKENING

The evidence for the protolaccolith spreading to its full diameter as a thin sill was first presented by Davis (1925). I have restated the geologic evidence in assumption 6 in Chapter II. The evidence for laccoliths spreading as thin sills appears conclusive. Regardless of the diameter of the laccolith, the magma spread initially as a sill <30 m thick.

The total load, F, on the roof of the protolaccolith increases as the square of the radius, r^2. The resistance to bending or shear, Q, is linear in r. Hence, at some radius, a, the load, F, will exceed the resistance to roof deformation, Q. There are a variety of factors which determine the final radius, a.

Diameter versus depth

Gilbert (1877) suggested that the diameter of the laccolith is related to the depth of intrusion. In a theoretical analysis, Pollard and Johnson (1973; see also Jackson and Pollard, 1988) suggested that the diameter of a laccolith is approximately related to the effective thickness of the overburden, z_e, by the following:

$$2a \sim 3z_e \qquad (8)$$

However, the field evidence presented in Appendix B does not provide any simple relation between depth of intrusion and diameter of laccoliths. One possibility for the lack of correlation between depth of intrusion and diameter is a systematic error in the observations. I have noted above that laccoliths commonly result from multilevel intrusions. The field observer usually has little information about the total amount of erosion since emplacement, or what level of the intrusion is presently exposed. Further, even if depth of erosion can be estimated, the relation between depth and diameter will not be as apparent for a Christmas-tree laccolith as for a punched type. Thus, it will be extremely difficult to find a correlation between depth of intrusion and diameter of Christmas-tree laccoliths from field

evidence, and no simple relation such as suggested by equation 8 can be verified.

Depth versus diameter relations are examined further in the section on theoretical models of laccoliths.

Rate of loading

In any problem in rock mechanics, the question of the strain rate is important inasmuch as both the strength and deformation mode of rocks is time dependent (Jaeger and Cook, 1976). Thus, to produce a reasonable model for the growth of laccoliths, an estimate is required for the rate of loading. Should the loading be very slow, time-dependent behavior of the overburden may become an important deformation mechanism, and the analysis more complex. However, time-dependent behavior can be ignored for fast loading, and elastic behavior will dominate until the loads reach the yield strength of the overburden.

An estimate of the loading rate can be made by calculating the time it would take for the sill to solidify. The estimate suffers the weakness that the magma is assumed to be stationary; in fact, new magma is constantly being added into the system. However, an estimate of strain rate that is accurate to plus or minus an order of magnitude is sufficient (Jaeger and Cook, 1976).

The time required for complete solidification of a sill has been calculated by Jaeger (1957, 1959, 1968). The time required for complete solidification must be considerably longer than the intrusion time. Savage (1974) has calculated the time for solidification of the Trachyte Mesa intrusion in the Henry Mountains, Utah. He calculates that, for a sheet intrusion 10 m thick, instantaneously emplaced at 1,000°C, in wet country rock with a porosity of 0.50, the time for complete solidification is 0.7 years. For zero porosity, the solidification time is 0.8 years. For a sheet 30 m thick, under the same conditions, the times to complete solidification are 6 and 7 years respectively. Assuming that a representative laccolith has a radius of 1 km, then the mean magma velocity in

45

the sill must be $>>5 \times 10^{-5}$ m s^{-1} for a sill 10 m thick if the magma is to spread to the margins before freezing. The mean magma velocity for a 30-m-thick sill with the same radius must be $>>5 \times 10^{-6}$ m s^{-1}. Sills with larger radii would require proportionately faster magma velocities. Thus, these velocity estimates are minimum values.

The further the laccolith spreads from the feeder dike, the cooler the magma must become due to contact with the cold country rock at the sill boundaries. As the magma cools, it becomes more viscous, and spreads ever more slowly. Hence, calculation of the loading rate based on time for solidification must be a minimum. The spreading rate may be increased if the laccolith has multiple feeder dikes, as observed by Gould (1926a) beneath Mount Mellenthin in the La Sal Mountains, Utah, or at Bee Mountain, near Study Butte, Texas.

Delaney and Pollard (1981, p. 51–53) review flow and dike propagation rates for basic magma and conclude that propagation rates of 0.1 to 1.0 m s^{-1} are to be expected with basic magmas. Spence and Turcotte (1985) calculate that a crack 2 m wide, with a magma viscosity of 10^2 Pa s, must have a velocity of about 0.25 m s^{-1} if the magma is to traverse the lithosphere without solidifying. They also point out (their Fig. 6) that as velocity decreases, the crack width must increase if the magma is to traverse the lithosphere without solidifying. Silicic magmas, because of their greater viscosity, probably have propagation rates 10^{-1} to 10^{-3} slower than basic magmas. Turcotte and Emerman (1985) require magma velocities on the order of a meter per second to prevent solidification during magma transport through the cold lithosphere. The models of Spera and others (1985) imply magma velocities of 1 to 20 m s^{-1}.

Even the slowest estimate for magma spreading represents fast loading as defined by Jaeger and Cook (1976). Thus, I can ignore time-dependent behavior, and the country rock deformation in the epizone can be treated as elastic-plastic.

From these estimates, the total time for intrusion of protolaccoliths is generally less than 1 year, and almost certainly less than 10 years. The laccolith itself may continue to grow vertically for a much longer period, since it is a much thicker body and will not freeze for a much longer period. However, an inescapable conclusion is that laccoliths are emplaced instantaneously on a geologic time scale.

Direct observation of laccolith growth

The above estimate of time for intrusion is in accord with the only direct observations of the growth of laccoliths of which I am aware. The growth of two laccoliths was observed at Usu-san, between Sapporo and Hakodate on Hokkaido Island, Japan. In 1910, a laccolith 2.7 km long and >0.6 km wide was elevated 155 m (~ 0.2 km^3 of magma) within a time period of 100 days (Omori, 1911a, 1911b; Bailey, 1912; Oinouye, 1917) before deflating to a final elevation of 99 m. In late May of 1944, a second laccolith, Syowa Sinzan, 1 km long by 0.8 km wide, began to grow (Minakami and others, 1951; Yagi, 1953). The develop-

ment of this dome was sketched by the postmaster in a nearby village. His remarkable diagram, which clearly shows the development of the laccolith as a function of time, is reproduced in Figure 34. Syowa Sinzan continued to grow for about 9 months to a total height of 170 to 200 m (~ 0.1 km^3 of magma). In contrast to the laccolith formed in 1910, the 1944 intrusion developed a volcanic neck of very viscous dacite magma, which pierced the roof. The late-stage growth of this neck is clearly shown in Figure 34. The mechanics for the formation of such late-stage necks is illustrated in Figure 19, and an eroded remnant is shown in Figure 20. The rapid growth of similar domes is reviewed by Williams and McBirney (1979, p. 190–196).

Pressure gradient in the protolaccolith

Another variable to be considered is the pressure gradient, dP_d/dr, between the feeder dike and the margin of the sill forming the protolaccolith. Johnson and Pollard (1973) have presented the basic equations. My purpose is to put bounds on the possible values of dP_d/dr. An upper bound can be easily established. For an estimated maximum magma driving pressure of about 65 MPa (Fig. 29), a pressure gradient of 65 kPa m^{-1} would give zero pressure at the margin of a laccolith that grew to a radius of 1 km. Since laccoliths of such dimensions are relatively common, the pressure gradient must be $<<65$ kPa m^{-1}. The pressure gradient can be further constrained by the observation that the deformation associated with laccoliths is concentrated at the margins. The protolaccolith begins to grow vertically only when the diameter has increased to the point where the total load on the roof exceeds the failure strength of the roof rock. If a significant pressure gradient exists, say >1 kPa m^{-1}, then the deflection of the roof should be greatest over the feeder dike because the magma pressure would be maximum there.

Using linear elasticity theory, Pollard and Muller (1976) have examined the effect of a pressure gradient due to a magma with a finite strength (they estimate a yield strength of 10^3 Pa) in a sill in Theater Canyon on the Mount Holmes laccolith, Utah. The effects amount to, at most, a 1- to 2-m difference in roof deflection over a 1-km length of the sill.

Field observations indicate that, in general, the roofs of laccoliths are relatively flat in all but the most extreme uplifts. No field evidence is available to suggest a significant pressure gradient between the feeder dike and the distal margin of the protolaccolith. The pressure gradient, dP_d/dr, can thus be assumed to be approximately zero.

It is probable that large laccoliths are fed by more than one feeder dike, although field evidence is scanty. Multiple feeder dikes would act to equalize pressures within the protolaccolith and thus strengthen my assumption of a pressure gradient near zero. Zero pressure gradient ($dP_d/dr \simeq 0$) implies the magma behaves as a static Newtonian fluid.

In estimating the pressure gradient, I have assumed the entire sill advances at a mean velocity of $\leqslant 1$ m s^{-1}. The assumed mean velocity does not appear to be physically realistic, however.

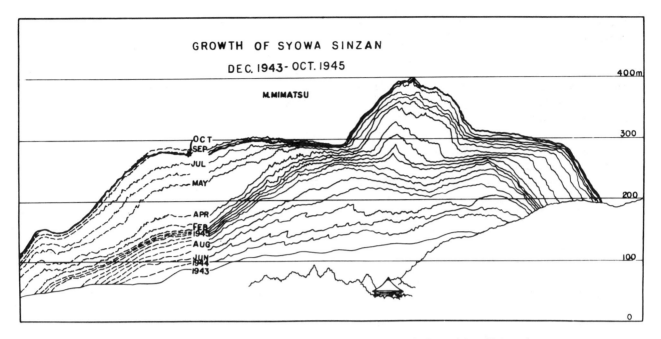

Figure 34. The vertical development of Syowa Sinzan (New Roof Mountain) at Usu-san between Sapporo and Hakodate, Hakkaido Island, Japan. Adapted from a guidebook by Masao Mimatsu, who sketched the growth of the laccolith from his post office. The laccolith is approximately 1 km long. The heavier line indicated by the arrow is the present outline since some slight subsidence in October, 1945. This sketch was originally published by Minakami and others (1951). Drape folding is obvious on the right side of the sketch. Note the formation of the late stage (after February, 1945) neck in what was a graben area. Devils Tower (Fig. 20) is thought to have formed in similar fashion. Volcanic necks or bosses on the roofs of laccoliths appear to be common.

As pointed out in stage 2, the magma advances as a series of fingers that extend in front of the radius of the main sill. The pressure gradient within the fingers and within the magma channels that feed the fingers may be large. However, if the instability hypothesis of Pollard and others (1975) is valid, then by definition, the fingers are advancing on only a fraction of the periphery at any given time. The remainder of the periphery is not advancing, or advancing relatively slowly. Thus, most of the sill forming the protolaccolith is nearly static at any given instant. The channels feeding the advancing fingers, where a substantial pressure gradient may exist, will be areally insignificant. The assumption of static Newtonian behavior is thus further reinforced.

An important point brought out by Pollard and others (1975) is that the fingers of the advancing sill extend beyond the margin of the laccolith at Shonkin Sag, Montana, by up to 300 m. There the laccolith roof did not fail at the outer periphery of the sill, as represented by the tips of the advancing fingers. Instead, failure occurred at some point behind the finger tips, presumably at the margin of the sill where the instabilities that form the fingers develop. Roof failure should occur over the minimum possible surface. For punched laccoliths, where failure is by shear at the periphery, the failure surface should then tend toward a cylinder. Thus, an irregularly shaped protolaccolith may develop into a radially symmetric laccolith. This effect will be much

less evident for laccoliths whose roofs deform ductilely. Field observations bear this out, since punched laccoliths are more commonly almost circular.

When the total load on the roof exceeds the failure strength of the overburden, the protolaccolith enters a phase of vertical growth. Because more than one sill usually loads the roof, no simple relationship can be used to predict what the radius at failure will be. Additional factors affecting the radius of the protolaccolith at the point of roof failure are examined subsequently with the theoretical models.

STAGE FOUR. LARGE-SCALE DEFORMATION OF THE OVERBURDEN BY THICKENING OF THE INTRUSION

The processes associated with stages one through three appear to be common to all laccolithic intrusions. This is not true of the fourth stage, as witness the diverse final configurations assumed by different laccolithic bodies.

In order to model the deformation associated with the growth, I have devised a classification scheme suggesting that two distinct shapes—punched and Christmas tree (Fig. 8)—represent end members of the series of growth modes. The entire series of

growth modes would represent all the various observed configurations of laccoliths.

The justification for assuming such a classification scheme is
to devise a mechanical basis for the observed shapes of laccoliths.
As discussed previously, variations of the magma variables that
might cause diverse shapes of laccoliths have been shown by
reasonable arguments to be insignificant if I examine laccoliths
within the same group. Therefore, it is physically unrealistic to
relate the observed variation in the shapes of laccoliths with
variations in magma rheology.

It would still be possible to explain the wide variation in the
shape of laccoliths by variations in the mechanical properties of
the overburden. However, a wide variation in the shape of laccoliths within the same group is commonly observed. It is unrea-
sonable to assume that the mechanical properties of the
overburden vary greatly within the area of a single laccolith
group when the same formations are found throughout the group
and the intrusions occur at roughly the same depth.

Some or all of the variation in configuration observed within
a group may be attributed to interference of nearby intrusions
(Fig. 9). This does not invalidate my premise of a series of growth
modes with distinct end members. Some interference with the
advance of the magma may be necessary to obtain Christmas-tree
laccoliths in a group where punched laccoliths are also found, or
vice versa.

Since punched laccoliths are conceptually and mechanically
the simplest, it is convenient to attempt a description of punched
laccoliths first.

PUNCHED LACCOLITHS

Description

Punched laccoliths form in the epizone when a single sill
dominates the country rock loading. In the field, punched laccoliths are recognizable by their flat tops, peripheral faults, and steep
or vertical sides. Examples are Table Mountain in the Henry
Mountains (cover) and Mount Peale (Fig. 17) in the La Sal
Mountains, Utah. Many other examples exist, and punched laccoliths were first described using a mechanical model by Gilbert
(1877). Iddings (1898) referred to this type of intrusion as a
bysmalith, a term I favor abandoning, since it is associated in the
literature with mechanically untenable arguments regarding the
role of magma viscosity.

In plan view, punched laccoliths are roughly cylindrical
with a shear zone forming a fault at the periphery. The roof
contacts are almost flat, even where uplift has been in excess of
1 km. However, some distortion of the roof is usually inevitable
with such large displacements (i.e., the laccolith will not form a
perfect right circular cylinder). The peripheral shear zone
(Fig. 17) cannot be traced laterally any significant distance into
the roof, nor into the surrounding country rock. The sill that
formed the protolaccolith may, in places, extend beyond the
periphery at the floor of the laccolith, though this is rarely visible.

Field observations

Any model of punched laccoliths is constrained by the following field observations:

1. Initial failure of the roof rock is by shear at the periphery.
This shear may be plastic flow, or brittle fracture. At least some
plastic flow appears to have occurred prior to failure at the margins of all punched laccoliths I have seen (e.g., Fig. 35). Savage
(1974) and Koch and others (1981) have treated the problem
quantitatively for Trachyte Mesa, Utah (Fig. 35), and Koch and
others (1981) have extended the analysis to Shonkin Sag,
Montana.

2. Distortion of the elevated roof rock is minimal away
from the periphery (e.g., see Fig. 9).

3. The country rock is undisturbed a short distance away
from the peripheral fault (Fig. 17). Occasionally dikes may
branch off from the flank of the intruding magma, but these dikes
usually die out within a few meters. Subsidiary sills may form at
distinct horizons within the country rock but only rarely are they
of any lateral extent. However, some laccoliths appear to have
used one of these subsidiary sills to extend laterally, assuming an
intermediate shape not treated here.

4. The thickness of the laccolith is equal to the total deflection of the roof.

5. The total deflection can be determined either by direct
measurement or by calculation of the structural relief of the roof
rocks.

6. Diameter can be determined by field observation.

7. In map view, punched laccoliths are commonly almost
circular.

8. Circular laccoliths commonly have a greater thickness
than elliptical laccoliths (Appendix B).

9. Depth of intrusion can be determined, at least approximately, from the stratigraphic section if the roof contact is
exposed.

Inferences

From the field observations, I make the following
inferences:

1. The roof was once in contact with the floor directly
below (i.e., the roof has not translated laterally).

2. As discussed in stage three, from observations that failure
is by shear at the periphery, and the laccolith remains flat topped,
I infer that the pressure gradient in the protolaccolith is approximately zero ($dP_d/dr < 1$ kPa m^{-1}). If $dP_d/dr \sim 0$, the magma
can be treated as a static Newtonian fluid.

3. Shear stress at the boundary of a static Newtonian fluid is
zero. Therefore, shear stress at the magma–country rock interface
is zero in the plane of the intrusion. For $dP_d/dr \sim 0$, the magma
acts as a perfectly lubricated, flat-ended punch. In the plane of
intrusion, but for a radius greater than the intrusion, the shear
stress is zero, since one principal stress axis due to body forces is
vertical. The plane of intrusion is sensibly parallel to the earth's

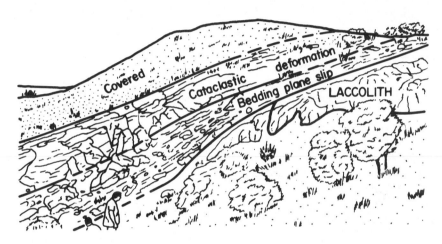

Figure 35. Plastic deformation over the flank of Trachyte Mesa laccolith, Utah. The laccolith stopped growing just at the point where shear fractures were developing in the periphery. See discussions in Johnson and Pollard (1973, their Fig. 8), Savage (1974), and Koch and others (1981). The view is looking east on the northwest flank of the laccolith.

surface. Therefore, this plane must contain the remaining two principal stress axes.

4. The overburden behaves initially as an elastic plate (see discussion in assumption 6 in Chapter II). As the diameter of the sill increases, the total load, F, on the roof increases as $P_t \pi r^2$, and plastic regions develop in the overburden around the periphery.

Boundary conditions

From these inferences and observations, I derive the boundary conditions for the intrusion at the onset of stage four.

1. There is no shear stress ($\tau_{r\theta} = 0$) at the magma–country rock interface in the plane of the intrusion.

2. The normal stress on the roof is given by equation 6, and

$2L_c$ is determined by reference to the table in Figure 28, where $\bar{D}$ can be measured, and P_d is constant ($dP_d/dr \sim 0$) across the base of the roof.

3. As the radius of the sill forming the protolaccolith increases, the total load, F, on the roof above the sill is given by:

$$F = P_t \pi r^2. \qquad (9)$$

4. The force required to separate the beds perpendicular to the plane of the intrusion is approximately equal to the weight of the overburden ($\tau_0 \sim 0$), that is, the intrusion is initially along a parting surface, or freely slipping joint, with an effective cohesive strength near zero parallel to the earth's surface. Suitable parting surfaces are assumed to always exist within the vertical range of neutral buoyancy.

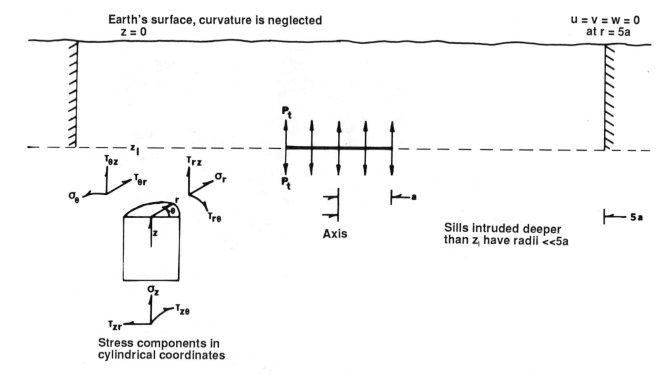

Figure 36. Theoretical model of a punched laccolith. a is the radius, r, at the point of roof failure.

5. Failure of the roof is by shear at the periphery.

6. Vertical displacement of the roof is within the elastic limits of deformation of the overburden (w < 30 m) until peripheral failure occurs. Plastic zones are progressively developed in the roof rock around the periphery as spreading progresses and the load increases.

7. Sills intruded deeper than z_1 (Fig. 36) have radii much less than that of the protolaccolith.

The boundary conditions are stated explicitly in Table 1, and the theoretical model is shown in Figure 36. Saint Venant's principle is used to justify the assumption that at a distance of five times the final radius, a, of the laccolith, the effects of the intrusion are negligible. The laccolith is assumed to be axisymmetric to simplify analysis; this assumption matches field observation reasonably well.

I have shown above that the loading of the roof is sufficiently fast that time-dependent behavior of the country rock can be ignored. The country rock can be treated as a continuum, as discussed in the assumptions. That the loading by the intrusion eventually exceeded the yield strength of the country rock is obvious from the field relations. Thus, the most appropriate rheology for the overburden is elastic-plastic.

Plastic deformation results when stress levels reach some yield condition. The most widely used yield condition is that of von Mises:

$$J_2 - k^2 \geqslant 0 \qquad (10)$$

where I assume the roof rock work hardens beyond the yield point. However, rock usually exhibits behavior transitional between elastic and plastic at stress levels near the yield point. Also, in the following finite element analyses, it is more convenient and accurate to define the behavior based on the total volumetric strain. For the models, I have used the following definition of material behavior. The rock is considered to be brittle (elastic) if the volumetric strain is <0.005; the rock is transitional for a volumetric strain $\geqslant 0.005$ and <0.010; and the rock is ductile (plastic) if the volumetric strain is $\geqslant 0.010$.

Prandtl punch. The problem as defined can be approximated under conditions of plane strain as the indentation of a finite body by a flat, lubricated, rigid punch. During the period of protolaccolith growth, when the radius, r, is small compared with the depth of emplacement, z_1, the overburden may be treated as a semi-infinite half space. The initial problem then has a solution similar to that given by Prandtl (1920; Hill, 1949; as quoted by Prager and Hodge, 1951) for incipient plastic flow. The Prandtl (1920) model is shown in Figure 37. The x axis in Figure 37 is assumed to correspond to the plane of intrusion. Plastic regions will begin to form at A and B (Fig. 37) as soon as a load is applied to the punch. Use of the von Mises yield criterion implies that the material between the local plastic regions does not initially indent. Only when a plastic region extends across the entire bottom of the punch does indentation become possible. No indentation occurs until the load applied to the punch has reached the value necessary to produce such an extended plastic region

TABLE 1. BOUNDARY CONDITIONS FOR MODELS OF PUNCHED AND CHRISTMAS-TREE LACCOLITHS
Punched Laccoliths

Boundary Condition	Locality (see Fig. 36)
$u = v = w = 0$	$r = 5a$
$\sigma_z = P_t = P_d + P_l$	$r \leq a, z = z_l$
$\sigma_z = -\lvert \bar{\rho}gz \rvert$	$r > a, z \leq z_l$
$\sigma_z = \tau_{zr} = \tau_{z\theta} = 0$	$r \leq 5a, z = 0$
$\tau_{zr} = \tau_{z\theta} = 0$	$r \leq 5a, z = z_l$
$\sigma_r = \sigma_\theta = \lvert \bar{\rho}gz \rvert [\nu/(1-\nu)]$	$r = 5a, z \leq z_l$
$\tau_{rz} = \tau_{r\theta} = 0$	$r = 5a, z \leq z_l$

Christmas-tree Laccoliths

Boundary Condition	Locality (see Fig. 40)
$u = v = w = 0$	$\lvert x \rvert = \lvert y \rvert = \lvert 5a_n \rvert$
$\sigma_z (i^{th} \text{ sill}) = P_t = P_d + P_l$	$\lvert x_i \rvert = \lvert y_i \rvert \leq \lvert a_i \rvert, z = z_i$
$\sigma_z = -\lvert \bar{\rho}gz \rvert$	$\lvert x_i \rvert = \lvert y_i \rvert > \lvert a_i \rvert, z \leq z_n$
$\sigma_z = \tau_{xy} = \tau_{zy} = 0$	$\lvert x_0 \rvert = \lvert y_0 \rvert \leq \lvert 5a_n \rvert, z = 0$
$\tau_{zx} = \tau_{zy} = 0$	$\lvert x_i \rvert = \lvert y_i \rvert \leq \lvert 5a_n \rvert, z = z_i$
$\sigma_x = \sigma_y = \lvert \bar{\rho}gz \rvert [\nu/(1 - \nu)]$	$\lvert x \rvert = \lvert y \rvert \leq \lvert 5a_n \rvert, z \leq z_n$
$\tau_{xz} = \tau_{xy} = \tau_{yz} = \tau_{yx} = 0$	$\lvert x \rvert = \lvert y \rvert = \lvert 5a_n \rvert, z \leq z_n$

(Prager and Hodge, 1951). As the radius, and thus the load imposed by the protolaccolith, increases, the overburden looks less like an infinite half space. Since in a protolaccolith the radius of the punch changes together with the load, and the half space above the intrusion cannot be treated as infinite (i.e., the problem is three dimensional), an analytic solution of the boundary value problem is not easily obtained. Savage (1974) and Koch and others (1981) have solved the problem in two dimensions for the flexure prior to shear failure observed at Trachyte Mesa laccolith, Utah (Fig. 35), and Shonkin Sag laccolith, Montana. I have used the finite element method to accommodate the three-dimensional nature of the problem. The numerical analysis is treated subsequently in the section on theoretical models.

Limital thickness

It is possible to treat Gilbert's concept of a "limital thickness", w_o, for epizonal laccoliths analytically. Assuming the roof remains intact over the growing laccolith, Gilbert (1877, p. 85) proposed, "When the sum of the weights of the cover and laccolite equals the total pressure of the intrusive lava, uplift ceases, and the maximum . . . thickness is obtained."

Semi-infinite half space

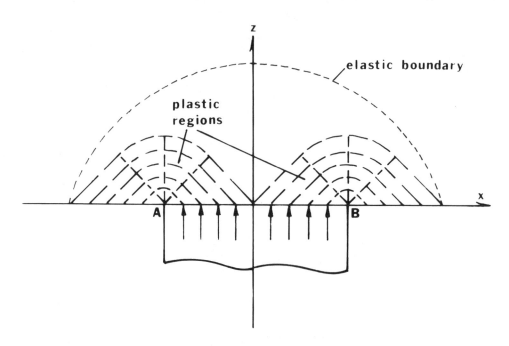

Figure 37. Stress field above a lubricated, flat, rigid punch. After Prandtl (1920). Redrawn from Prager and Hodge (1951). Dashed lines indicate initial shear lines or slip planes.

From previous discussion, the roof of a punched laccolith moves up as a nearly rigid cylindrical plate on top of the magma until hydrostatic pressure balances the stress drop that occurs at failure around the perimeter of the cylinder. This condition can be stated (Corry, 1972) as

$$w_o = (2kz_1) / (\rho'ga). \qquad (11)$$

Equation 11 can be justified by reference to the model shown in Figure 38. At equilibrium, and just prior to failure

$$\Sigma(F - Q) = 0 \qquad (12)$$

and

$$\begin{aligned} F &= P_t\pi r^2 \\ Q &= 2\pi rkz_1 + \pi r^2\bar{\rho}gz_1. \end{aligned} \qquad (13)$$

At equilibrium

$$P_t\pi r^2 = 2\pi rkz_1 + \pi r^2\bar{\rho}gz_1 \qquad (14)$$

and therefore

$$P_t = [2kz_1) / r] + \bar{\rho}gz_1. \qquad (15)$$

(The variables for these equations are defined in Appendix C). As the sill continues to spread, the radius, r, increases by some small increment, Δr, to a, which increases the total load, F, on the roof to the point of failure. After failure, the shear strength, k, of the overburden at the perimeter goes to zero. Assuming that the magma driving pressure, P_d, remains constant at the outlet of the feeder dike, and neglecting friction on the cylinder walls, the thickness of the laccolith will increase until the pressure is again balanced by the overburden weight plus the weight of the magma, at which point

$$P_t = \rho'gw_0 + \bar{\rho}gz_1 \qquad (16)$$

Using the right-hand side of equations 15 and 16, and solving for w_0, I obtain

$$w_o = (2kz_1) / (\rho'ga). \qquad (11)$$

In the derivation, I assumed that the downward-acting force, Q, will be reduced an amount equal to k when the strength of the overburden is exceeded. The problem can also be solved by letting k represent a stress drop associated with the transition from static to dynamic friction, or a shear strength to dynamic friction stress drop. The process has been modeled by Ramberg (1981), and his results are shown in Figure 39. Regardless of the

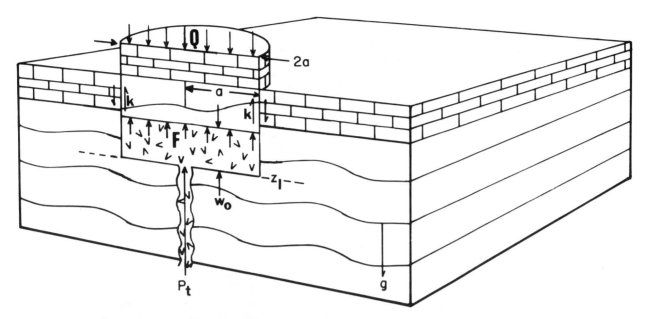

Figure 38. Illustration of forces required to move a cylinder vertically through a resisting medium. The problem is approximately that of a flat-ended punch problem. F is the force exerted upward by the intrusion and Q is the downward-acting force of the overburden. a is the radius, r, at failure. w_o is the limital thickness and k is the shear strength of the overburden. z is the level of intrusion and g is the acceleration due to gravity. k is the yield limit in simple shear, or shear strength.

assumptions, it is clear that k represents the failure strength. Since a circular punch approximates pure shear at the perimeter of the punch, if k can be calculated, the bulk shear strength of the roof rock can be determined.

In obtaining equation 11, I assumed that the magma pressure, P_t, remains constant. For large laccoliths, and in the final stages of laccolith growth, the assumption is probably not valid. If the magma pressure, P_t, does vary as the laccolith grows, it is probable that the change will be a drop in pressure. Equation 11 must then be written as the inequality

$$w_o \leqslant (2kz_1) / (\rho'ga). \qquad (17)$$

All of the parameters of equation 17 are subject to field determination except k. It is then straightforward to write

$$k \geqslant (w_o\rho'ga) / (2z_1) \qquad (18)$$

Thus, a reasonable minimum value for the yield limit in simple shear, k, can be determined for a laccolith group by using the dimensions of the thickest and largest diameter punched laccolith within the group. An example would be Table Mountain in the Henry Mountains, Utah. Table Mountain (cover and Appendix B) has a radius, a, of 1,050 m, a thickness, w_o, of about 810 m, and a depth of intrusion, z_1, of 1,100 m (Hunt and others, 1953).

Assuming a magma density of 2,500 kg m^{-3}, the shear strength, k, of the country rock must have been $\geqslant$9.5 MPa.

An examination of the possible range for the limiting thickness suggests a maximum of approximately 2 km for epizonal laccoliths. A lopolith may grow thicker by simultaneously depressing its floor. The derivation contains the assumption that the laccolith roof is not breached, and the overburden weight is treated as a constant. Suppose, however, that the weight of the roof decreases as the thickness of the laccolith increases. In that instance the laccolith thickness becomes indeterminate. Consider a punched laccolith with an overburden composed of alternating layers of limestone and water-saturated clays or shales. Little imagination is needed to see that such an unstable mass rising up to 2 km in the air could result in the development of gravity slump blocks that might have unroofed the laccolith as fast as it grew. The laccolith might then have continued to grow until the magma reached the surface.

CHRISTMAS-TREE LACCOLITHS

Description

Christmas-tree laccoliths form by ductile deformation of their roof. Gilbert (1877) called these "compound laccoliths" and clearly recognized that some laccoliths are multileveled. In the field, Christmas-tree laccoliths are recognizable by their rounded dome appearance. They do not have peripheral faults, and the

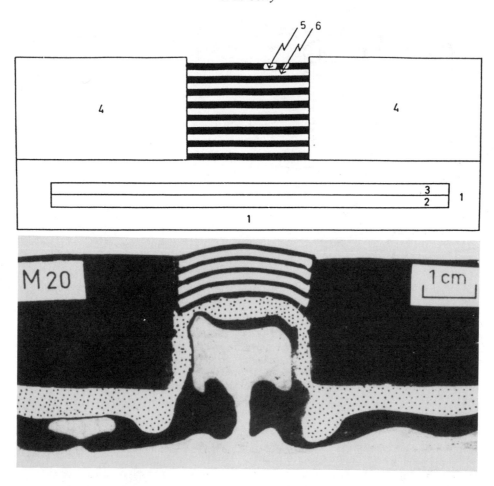

Figure 39. Centrifuge model simulating the conditions for Gilbert's concept of limital thickness.
  View above shows the initial structure of the model. 1 - white painters putty, 2 - pure silicone putty, $\rho = 1{,}140$ kg m^{-3}, 3 - silicone putty with magnetite powder, $\rho = 1{,}330$ kg m^{-3}, 4 - massive modelling clay, 5 - dark and 6 - white modelling clay, ~1.5 mm sheets. The model is 9.5 cm long.
  View below shows model after running for 7 minutes at accelerations varying between 1,000 g and 1,500 g. Continued run in the centrifige did not elevate the block further, indicating the limital thickness had been reached. After Ramberg (1981, p. 292).

overlying rocks are continuous across the laccolith where not removed by erosion. The extension over the dome has been accommodated by ductile thinning of the roof rock. A crestal graben forms as uplift continues. Well-exposed examples of Christmas-tree laccoliths are uncommon. The characteristic Christmas-tree sill arrangement was originally described in cross section from a laccolith by the Moldau River in Czechoslovakia by Kettner (1914; reproduced in Hunt and others, 1953). Excellent cross sections of a Christmas-tree laccolith are exposed in the Christmas Mountains (Fig. 11) in west Texas and in the Tenmile district, Colorado (Fig. 12), as mapped by Emmons (1898). The best example I have seen of the smooth dome over a Christmas-tree laccolith is Green Mountain near Sundance, Wyoming, shown in Figure 10.

Christmas-tree laccoliths may vary from circular to elliptical in map view. It is unusual for the dimension of the major axis to exceed the minor axis by more than a factor of 5 and a factor of $\leqslant 2$ is more common.

Christmas-tree laccoliths are also flat-roofed over the center, though it is often difficult to establish this in the field. The flat central area is frequently small and distorted by a crestal graben. Christmas-tree laccoliths are as distinctive as punched, though not usually as spectacular.

I have previously argued that the physical properties of the magma which forms all the laccoliths within a group are the same. No obvious differences in mechanical properties of the overburden are seen within a group. Yet both punched and Christmas-tree laccoliths coexist within the same group, often in close proximity, and with virtually identical igneous rock types.

The stress concentrations at the boundaries associated with

the margins of the intrusion in punched laccoliths, where a single sill is dominant, *must* result in shear failure at the periphery in epizonal rocks. The technique of using a punch to produce shear is a standard experimental technique (Jaeger, 1962; Maurer, 1965; Jaeger and Cook, 1976). Shear failure on a large scale is demonstrated by punched laccoliths. Thus, a principal task is understanding why Christmas-tree laccoliths do not develop peripheral faults. The solution cannot call on variations in magma or overburden properties to account for the lack of a fault, because that would contradict the field evidence. Thus, I propose that the critical factor must be the loading condition imposed on the overburden.

Sills are commonly found closely associated with laccoliths. Usually these sills are of limited extent. In punched laccoliths I have imposed the condition that these sills are of small radius and, thus, mechanically insignificant. However, if the deeper sills grow faster than the higher sills, for whatever reason (e.g., interference from earlier intrusions or small variations in local lithology), then the loading condition on the overburden is radically different from the loading for a punched laccolith. If the deepest sill has the largest radius, but higher sills have substantial and progressively decreasing radii (i.e., a Christmas-tree arrangement), then a Christmas-tree laccolith may develop. Any intermediate sill configuration could give a form intermediate between a punched and a Christmas-tree laccolith.

Since each sill is pressurized in a radially progressive sequence, plastic zones will develop above each sill as the intrusion progresses. As sill diameters continue to increase, the plastic zones of deeper, wider sills will merge with the plastic zones of shallower sills. When the plastic zones intercept the surface, the overburden is deflected upward. Since the country rock is in a plastic state, except for the area at the surface above the center of the shallowest sill, and no sharp discontinuities exist at the edges of the sills, a symmetrical dome is formed.

I assume that the vertical spacing between sills is sufficiently small that the plastic zones merge before the surface is affected by the intrusion. If this condition is not met, then an intermediate form between a punched and Christmas-tree laccolith is the probable result.

Field observations

The model for Christmas-tree laccoliths is constrained by the following field observations:

1. Beds are continuous across the laccolith and deform by a process analogous to drape folding, except that the deformation is circular. Unless the dome has deflected to the point that a crestal graben has formed, no faults with significant vertical (>1 m) offset have formed. A series of hinge zones, outlined by distinct fracture zones around the periphery, can be found. Deformation is largely limited to these hinge zones, in my experience, although bedding plane slip within the blocks is known to occur. The hinge zones act as pivots for the observed block rotations.

2. Peripheral faults have not formed. An example is shown in Figure 10.

3. With sufficient deflection, tensile stress develops in the crest of the dome and a crestal graben results.

4. Maximum deflection can be obtained from the structural relief if the roof is at least partially intact. Total deflection exceeds the thickness of any individual sill in the laccolith.

5. Diameter of the laccolith does not correlate with the diameter observed at the surface because the overburden is folded beyond the limits of the intrusion.

6. In map view, Christmas-tree laccoliths range from circular to elliptical. The dimension of the major axis seldom exceeds five times the minor axis.

7. Depth of intrusion of the various sills is indeterminate. Only by drilling does it appear possible to determine depths of the various layers of intrusion.

8. The sills are usually not all intruded simultaneously. Because of the limited contraction during solidification, the stress field around previous sills is frozen in.

Inferences

From the field observations I make the following inferences:

1. Continuity of the beds across a Christmas-tree laccolith, where vertical deflection may exceed a kilometer, requires plastic deformation.

2. From assumption 4 in Chapter II (i.e., all magmas in a group of laccoliths had similar physical properties), I infer that if in punched laccoliths the magma behaves as a static Newtonian fluid, then it must behave the same way in Christmas-tree laccoliths.

3. If shear stresses are zero in the plane of intrusion for punched laccoliths, then shear stresses are also zero on each level of intrusion for Christmas-tree laccoliths.

Boundary conditions

From these inferences and observations I derive the following boundary conditions for the onset of stage four. From assumption 4 (in Chapter II), and the argument for similar overburden properties within a group, I am constrained to conditions for Christmas-tree laccoliths that are essentially those for punched laccoliths:

1. There is no shear stress, $\tau_{xy} = 0$, at the magma–country rock interface in the planes of intrusion.

2. The normal stress on the roof of each sill is given by equation 6 (Chapter IV), and $2L_c$ is determined by reference to the table in Figure 29 (Chapter IV), where $\bar{D}$ is known. P_d is constant ($dP_d/dr \sim 0$) across the base of the roof of each sill.

3. To simplify analysis, I consider only circular laccoliths. Thus, the total load, F, on the roof increases as

$$F = \sum_{i=1}^{n} P_t \, \pi r_i^2 \tag{19}$$

and here P_t must be calculated at the ith level.

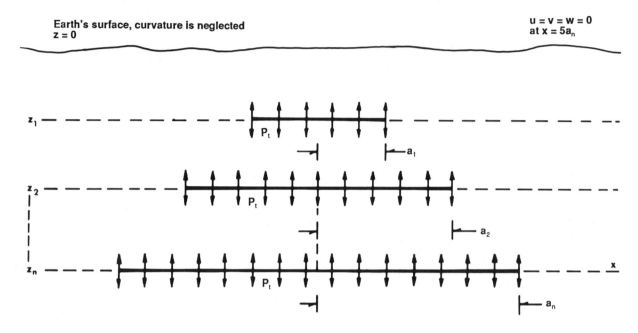

Figure 40. Theoretical model of a Christmas-tree laccolith. Sills intruded deeper than z_n have radii $\ll a_n$.

4. The force required to separate the beds in the planes of intrusion is approximately equal to the weight of the overburden ($\tau_0 \sim 0$).

5. Failure of the roof is characterized by plastic flow, presumably as some combination of bedding plane slip, cataclasis, and movement on preexisting fractures.

6. Vertical displacement of the roof is within the elastic limits of the country rock up to the point where roof failure occurs.

7. Plastic zones are progressively developed at the margins of each level of intrusion as spreading progresses and the load increases.

8. Sills intruded deeper than z_n (Fig. 40) have a radius $\ll a_n$.

The boundary conditions are stated explicitly in Table 1, and the theoretical model is shown in Figure 40. The boundary conditions are essentially the same as for punched laccoliths, with the exception that multiple loading is introduced, and the deepest sill has the largest diameter. From Saint Venant's principle I again assume that at a distance of five times the radius of the largest sill, a_n, the effects of the intrusion are negligible. The laccolith is assumed to be axisymmetric to simplify analysis, and this is justified for some Christmas-tree laccoliths (e.g., see Fig. 10), but by no means for all.

As in punched laccoliths, an elastic-plastic rheology is assumed. Since the difficulty of analysis is compounded over that of a punch problem, the boundary value problem is intractable. Again, the technique of nonlinear finite element analysis is used to obtain numerical solutions.

Theoretical analyses of the end members, punched and Christmas tree, based on the finite element method are presented in the following chapter. To a limited extent, analyses have also been made of the effects of differences in magma driving pressure and depth of intrusion.

CHAPTER VI

CONTINUUM SOLUTIONS FOR THE MECHANICS OF LACCOLITH GROWTH

THE FINITE ELEMENT METHOD

The finite element method is a general procedure for structural analysis in which a continuum with a complex shape subjected to arbitrary loading and boundary conditions can be solved numerically. The technique is commonly used for problems for which no convenient closed-form analytical solution is readily available. The structure is replaced by a finite number of elements interconnected at nodal points. The equations of continuum mechanics are then used to solve for the displacements, strains, and stresses within each element. By the law of superposition, the behavior of the total structure can be determined.

Forces acting on the actual structure, concentrated or distributed loads and body forces, are replaced by equivalent concentrated forces acting at the nodal points of the elements. Loading may be either conservative or nonconservative.

For nonlinear analysis, the isoparametric elements described by Zienkiewicz (1977) possess inherent advantages, and have been used exclusively in the present analysis. The Serendipity shape functions given by Zienkiewicz (1977) have been used to develop a mesh generation scheme.

The core of the present analysis is a program called MAG-GIE (Haisler and Corry, 1983). MAGGIE is a broad-based program for nonlinear analysis of static and dynamic structural problems based on isoparametric elements. Two- and three-dimensional elements for plane stress, plane strain, and axisymmetric analysis are included, as well as truss elements. The program calculates both geometric and material nonlinearities, and a wide range of nonlinear material models are included. The material model of interest in the present project is called the curve description model. In this material model the nonlinear material behavior is input as the bulk modulus, K, and shear modulus, G, as a function of the volumetric strain, e. However, what is available in the literature is an axial stress-strain curve for a discrete, intact sample of rock from which the tangent modulus, E_t, can be obtained at a given strain, ϵ. It is then necessary to find the bulk modulus by assuming a Poisson's ratio, $\nu(e) = 0.25$, which has

been but need not be, assumed constant at all strain levels. K(e) and G(e) for isotopic country rock are then determined from the relations:

$$K(e) = E_t(\epsilon) / 3[1 - 2\nu(e)] \qquad (20)$$

and

$$G(e) = E_t(\epsilon) / 2[1 + \nu(e)]. \qquad (21)$$

I have assumed a homogeneous overburden of Indiana limestone for all theoritical models, and the bulk and shear moduli used in all the analyses are presented on Plate VI. I have not yet investigated the effects of variations in overburden on the models.

Material nonlinearity

The program first calculates nodal displacements in global coordinates for a given load increment with a stiffness matrix based on the modulus at zero strain for the first step, and the strain at the previous load step in subsequent steps. The derivatives of the displacements are then used to obtain the new values for the strain at integration stations internal to each element. The new strain value is then used to find the current modulus from the bulk and shear modulus versus volumetric strain curve. The element stiffness matrix is assembled with the updated modulus, and current stresses are solved for. The global stiffness matrix is assembled by superposition, and the updated global load-displacement relations are obtained.

Geometric nonlinearity

The other type of nonlinearity that may occur is a geometric nonlinearity. MAGGIE includes total Lagrangian and updated Lagrangian formulations for geometric nonlinearity. My models have been run with material nonlinearities only, and with both material and geometric nonlinearities using the total Lagrangian technique. A total Lagrangian formulation makes the structure appear slightly stiffer than what an analysis allowing only for material nonlinearities would. However, no significant variation was found in the analysis of the same model run with and without geometric nonlinearities up to the point of roof failure. Using the total Lagrangian formulation allows for larger total deflections of the roof before the solution becomes unstable. Even larger deflections could be modeled using the updated Lagrangian formulation in MAGGIE, and deformation after roof failure can be modeled using dynamic analysis with nonconservative loading. While the latter capabilities are available, time has not yet permitted these analyses.

Integration stations

It is impossible to have both displacements and stress-strain simultaneously continuous across element boundaries for geometrically nonlinear analysis. This limitation is circumvented by keeping displacements continuous at the nodal points on the element boundaries and calculating strains and stresses at integration stations interior to the element.

Body forces

In a significant departure from most investigations, I have included body forces routinely in my models. Body forces are generated internally by the program as equivalent concentrated nodal forces for each element using a density of 2,670 kg m^{-3}. Only one model was run without body forces and, as discussed below, unrealistic behavior resulted.

Configuration of the models

All of the models are axisymmetric. Many laccoliths are in fact very nearly radially symmetric, so the assumption is in accord with many field observations. Saint Venant's principle was used to set displacements to zero at a distance five times the radius of the intrusion. Initial models were run with this condition, but it was found in practice that at a distance three times the radius of the intrusion, the displacements were effectively zero. Body forces were found to dominate the problem within a distance of two radii. Based on these findings, later models were run with the peripheral boundary moved to within two to three radii of the sill.

The application of body forces leads to considerations of the state of stress underground. Jaeger and Cook (1976, Chap. 14) and Engelder and Sbar (1984) have reviewed the problems in determining the state of stress due to gravity loading. While the problems are presently unresolved, at least one point seems clear: *The stress state in rocks at shallow depths in the crust is not, in general, hydrostatic.* The effects of a nonhydrostatic stress state are examined in model 6 for a punched laccolith.

The model boundary conditions are such that at the central axis, and at the outer periphery, only vertical translation is permitted. The models are pinned along the bottom. When body forces are applied with these boundary conditions, the horizontal stresses are related to the vertical stress by

$$\sigma_r = \sigma_\theta = |\bar{\rho}gz| \left\{ \nu/(1-\nu) \right\}. \tag{22}$$

To simulate the intrusion of the sill (i.e., stage three), a variant of the curve description model of MAGGIE was used. The assumption is made that the rock has no tensile strength. The program tests to see if the tensile stress exceeds the *in situ* stress due to gravity at each integration station in the element. If tensile stresses exceed the body forces, the element is "cracked" by reducing the modulus of the element in the direction of the tensile stress. The crack plane is therefore perpendicular to the tensile stress. Cracks are allowed to form in any direction, and elements may be cracked in more than one direction at the same time. To maintain numerical stability it is necessary to force an entire element to crack in the same direction if any one integration station cracks. The element remains cracked in this direction unless it closes (i.e., tensile stresses become less than the body forces).

The advance with time of an axisymmetric sill (protolaccolith) is represented in the theoretical models by a planar array of thin elements that are used to simulate the plane of intrusion. These thin elements are loaded internally with the calculated magma driving pressure, P_d, from stage one (Fig. 29) plus the overburden pressure, P_1, at the depth of intrusion. Since the sill exerts pressure down as well as up, loading the thin elements vertically places them in tension and causes them to crack in the horizontal direction. With the thin element cracked, the element modulus is reduced to approximately 10 from 10^9. The structure above the cracked element is thus vertically decoupled by the pressurized elements from the structure below. By loading the thin element at the central axis first, then the next element out from the axis, followed by loading each sequential thin element further out in its turn, the intrusion of a pressurized sill whose radius increases with time can be simulated.

The large aspect ratio of the thin elements used to simulate the intrusion plane causes some distortion of the stress trajectories within these elements. The distortion is not considered significant.

Care had to be taken to load the model sufficiently slowly that numerical stability was maintained since the analysis was quasi-static. There was thus no attempt to correlate the loading in the models with actual spreading rates of the sill(s) forming the protolaccolith.

In summary, the model includes a thin plane of axisymmetric elements that are used to simulate the plane of intrusion. These thin elements are sequentially loaded (i.e., pressurized), placing them in tension. When the tensile stresses in these thin elements exceed the body forces, the elements crack and the modulus in the thin element is reduced to near zero. The roof of the theoretical model laccolith is then sequentially loaded by compressive forces equivalent to a pressurized sill whose radius increases with time. The resolution of the method is obviously limited by the size of the elements used in the analysis. MAGGIE has no inherent limits on model size and allows in-core or out-of-core solutions for the stiffness matrix. However, for the sake of economy I have used fairly large elements (Plate VI), and for the same reason, no convergence tests were run.

THEORETICAL MODELS

Previous model studies

The early theoretical models presented by Gilbert (1877) have been discussed previously. Experimental models of the growth of laccoliths were first constructed by Howe (in Jaggar, 1901). MacCarthy (1925) performed some very interesting experiments, the illustrations of which led to prolonged debates with George M. Sowers. From these discussions my concept of punched and Christmas-tree end members for laccolith shapes evolved. The next experimental models I am aware of were done by Griggs (in Hurlburt, 1939). The modelling of laccoliths was then in abeyance until Pollard (1969) began his work on deformation of host rocks during the intrusion of laccoliths. He presented both a theoretical analysis based on linear elasticity, and physical models made of gelatin into which grease was injected. Pollard and Johnson have continued their modelling studies (Johnson, 1970; Johnson and Pollard, 1973; Pollard and Johnson, 1969, 1973; Pollard, 1972, 1973; Pollard and others, 1975; Pollard and Muller, 1976; Koch and others, 1981; Jackson and Pollard, 1988). I have used their studies as a departure point for my models, and my work would certainly have been a great deal more difficult without their efforts.

Stephansson and Berner (1971) introduced the finite element process to tectonic analysis in problems similar to the models I present. Ramberg (1970, 1981) and Simpson (1980) have studied models in centrifuges that are initially in density disequilibrium, and one of Ramberg's (1981) models is reproduced in Figure 39. Dieterich and Decker (1975) used the finite element method to model intrusions, but their analysis is limited to elastic behavior and body forces are not included.

Data display

For the theoretical models, I have made the simplifying assumption that laccoliths are radially symmetric. Thus, all models are axisymmetric, and cylindrical coordinates are used throughout. A two-dimensional vertical slice, from the central axis on the left to the periphery on the right, is displayed on the accompanying Plate VI. MAGGIE calculates stresses in the global r, z, and θ directions; the principal stresses in the r and z directions, and the principal direction of the maximum tensile stress relative to the r axis in the r–z plane; the volumetric strain; and whether cracking has occurred and, if so, the direction of the crack in the r–z plane and what type of crack has formed. All this information is printed out for each integration station in each element for each load step. Additionally, graphic output files are provided that can be displayed using the facilities of the MOVIE. BYU graphic program (Christiansen and Stephenson, 1983).

Because of the volume of output, a suitable method of displaying the output has been difficult to devise. Another problem occurs because of the inclusion of body forces that tend to obscure the intrusion stresses. While an ideal solution is unlikely, the format used in Plate VI provides a reasonable compromise.

No displacements have been shown for two reasons: (1) No displacement of more than a few meters were observed in the models before roof failure occurs and the deformation becomes unstable and execution is stopped. Such small displacements are not visible at the scale of the plots used in Plate VI. (2) Due to the inclusion of body forces in the problem, all initial displacements are in the –z direction. To obtain the displacements due to the intrusion it is necessary to store the displacements due to body forces and then subtract those values from the displacements after the intrusion. While it is feasible to do this, hand calculations for a few critical points did not indicate the displacements would be visible at the scale used in Plate VI; thus, it was not worthwhile.

Jaeger and Cook (1976) point out that the separation of the stress trajectories gives immediate insight to the stress intensity. Unfortunately, MOVIE.BYU does not plot stress trajectories and it was necessary to do the plots in Plate VI by hand. In the presence of body forces, any rotation of the maximum compressive stress from the vertical indicates significant perturbation of the stress field by magma loading. A limitation is that compressive stresses due to the intrusion also act vertically and locally may not rotate the stress trajectories. On Plate VI I have shown the maximum compressive stress direction by drawing a straight line through the integration station in the principal direction. Lines are only drawn where the deviation of the stress trajectory from vertical exceeds five degrees. Areas of high stress intensity are thus outlined by the stress trajectories. It is easily possible to contour the principal stresses, shear stress, and axial stress, but the inclusion of body forces makes much of the raw analyses ambiguous.

All models are shown at the point of instability; that is, roof failure has occurred because of loading applied during the previous load step. In the next load step, the solution becomes numerically unstable and the laccolith begins a period of unstable growth.

PUNCHED LACCOLITH MODELS

To form punched laccoliths I have proposed that the loading must approximate a circular punch and the loading is on a single surface above one dominant sill. The scheme is shown schematically in Figure 8, and the theoretical model is shown on Figure 36.

Model one

The depth of intrusion is 1,200 m, and the magma driving pressure, P_d, is 50 MPa. The roof failed in load step eleven, at which point two thin elements had been pressurized, simulating the spreading of the sill to 534 m, or a diameter of 1,068 m. As can be seen from Plate VI, the largest rotation of the stress trajectories, and hence the greatest stress intensity in the roof, occurs just above and around the end of the sill. The maximum shear stress correlates well with the amount of rotation of the stress trajectories for these models owing to the inclusion of body forces. Thus, the maximum shear stress in the roof of the model occurs at the integration station just above and to the left of the end of sill. At the integration station just above and to the right of the end of the sill, the shear stress is nearly two orders of magnitude less. Therefore, a vertical shear will develop at the end of the sill, since the differential shear stress between these integration stations is 13 MPa. As shown in the section on limiting thickness, the shear strength of the country rock in the Henry Mountains, Utah, was $\geqslant 9.5$ MPa. The total shear stress at the actual end of the sill is probably much greater than 13 MPa since the integration stations are a considerable distance from the end of the sill.

In calculating the differential shear stress, I have subtracted the shear stress at the integration station above and to the right of the end of the sill from the shear stress above and to the left of the end. An exact calculation of the maximum shear stress would require that shear stress be calculated at the node at the end of the sill. That cannot be done for geometrically nonlinear finite element analysis.

Model one (Plate VI) indicates that initial failure occurs at the surface and is due to tensile hoop stresses, σ_θ, which exceed the gravitational pressure to a depth of more than 100 m. Thus, in the model true tensile stresses develop at significant depths. Whether the surface fractures propagate downward or the corner of the sill shears upward is immaterial. What is clear is that the large stress at the outer edge of the intrusion will vertically shear the overburden as roof failure progresses.

The analogy with a Prandtl punch does not appear to carry through in the model. Plastic (transitional material behavior) zones do develop but do not extend upward to the degree indicated by the Prandtl punch analogy. The development of the tensile hoop stresses in the near surface overwhelms the development of the plastic zones when a free surface is close to the punch. However, plastic zones, evidenced by cataclasis, bedding plane slip, and drape folding (Fig. 35) may well be developed as a result of the hoop stresses. Koch and others (1981) have examined the problem of flexure at laccolithic margins using a two-dimensional approximation. If flexure occurs, the areas shown in extension on model one (Plate VI) may also represent plastic zones. The shear zone, which the model indicates will develop at the periphery of the sill, will also be represented by plastic deformation in most rocks at depths of one or more kilometers in the earth.

Model two

The depth of intrusion is kept at 1,200 m, the modulus of the rocks is the same, but the magma driving pressure, P_d, is lowered to 25 MPa. Roof failure occurred at the same radius, 534 m, as

for model one (Plate VI). The upper limit of the relation between radius and magma driving pressure has not been investigated. The lower limit is discussed in model three.

While the radius at failure is unchanged, the shear stress concentration between integration stations at the corner of the sill has decreased to 7 MPa, while the hoop stresses at the surface still exceed the gravity pressure. In fact, the model indicates the rock is in true tension to a depth of 300 m. The length of the area in extension at the surface has decreased a few hundred meters. Otherwise the deformation style has changed little for a change in driving pressure by a factor of 2. I would consider it unlikely that any distinction in the deformation style can be made in the field between laccoliths formed at magma driving pressures of 25 or 50 MPa.

Model three

Depth of intrusion remains at 1,200 m, and the rock modulus is unchanged. The magma driving pressure, P_d, is lowered to 10 MPa. Loading continued until the sill had a radius of 3.5 km without any observed roof failure. In fact, at this driving pressure the tensile stresses in the 10-m-thick elements used to simulate the sill was insufficient to crack the elements. Recall that the actual pressure, P_t, in the thin elements is equal to the magma driving pressure, P_d, plus the lithostatic pressure, P_l. The lithostatic pressure, P_l, is balanced by the internal pressure in the thin elements. The magma driving pressure, P_d, serves to deform the roof. The present models suggest that for roof failure to occur the following relationship must hold

$$2P_d \geqslant P_l \qquad (23)$$

at least for single-layer sills. Model three suggests that such sills can spread to great distances without roof failure, as Bradley (1965) predicted, if injected at sufficient depth that twice the magma driving pressure is less than the overburden pressure ($2P_d < P_l$). Mudge (1968, p. 325) suggests that one factor controlling the depth of intrusion is the overburden or lithostatic pressure, P_l. Laccoliths examined in his study intruded at depths between 1 and 2.5 km. He further suggests that the 2.5 km depth is near the maximum depth at which the magma driving pressure, P_d, is sufficient to lift the overburden. Using a maximum driving pressure of 65 MPa (as calculated, Fig. 29), and plugging into equation 23, suggests a maximum emplacement depth of approximately 5 km, at which it would be possible for felsic laccoliths to lift the overburden. For basic magmas, or superheated silicic magmas, or magmas with more contained volatiles, the driving pressure will be less, say on the order of 25 MPa, and the maximum depth of intrusion at which a laccolith can form will be about 2 km if equation 23 is valid. Laccolithic intrusions in the mesozone will tend to form lopoliths. As noted in model five, variations in the type of deformation will occur as depth increases to the point where the country rock is in a ductile regime due to body force loading.

It is possible that the total loading by multiple level sills may cause roof failure at magma driving pressures less than the overburden pressure. That possibility has not yet been investigated.

I have not included model three on Plate VI because no stress trajectories are rotated more than 5°, no areas go into extension, and no plastic zones develop. At the same load step and radius as models one and two when the roof failed, the differential stress between integration stations at the edge of the sill was only 500 kPa.

Model four

The depth of intrusion is still 1,200 m, and the rock modulus is unchanged. The magma driving pressure, P_d, is 25 MPa. To investigate the effect of body forces, I have deleted them in this model. For the imposed boundary conditions, model four is physically equivalent to including body forces but making the overburden incompressible (i.e., Poisson's ratio equals 0.50).

The loading of the roof continued until the radius of the sill had reached a radius of 3.5 km. No roof failure occurred. Plate VI shows the stress trajectories for model four at the same load step as for models one and two. The stress discontinuity at the corner of the sill is clearly shown by the stress trajectories, but the differential shear stress is only 1.2 MPa compared to 13 MPa and 7 MPa for models one and two, respectively. Model four suggests that any model of an intrusion process must include body forces.

The stress trajectories (Plate VI) are in excellent agreement with the theoretical ones obtained by Anderson (1938, his Figs. 1 and 3) except near the boundaries in the model. Anderson places his boundaries at infinity. The stress trajectories of the model (Plate VI) are, however, grossly different than those obtained by Pollard (1973) for his sheet intrusion model.

Model five

In model five the depth of intrusion has been increased to 2,400 m to investigate the effect of depth of intrusion versus diameter. The magma driving pressure is 50 MPa as in model 1. Body forces are again included. Roof failure occurred in load step 13, which corresponds to a radius of 1,067 m, or a diameter of 2,134 m. Thus, doubling the depth of intrusion has doubled the diameter of the laccolith. Pollard and Johnson (1973) had predicted that the transition from sill to laccolith would occur when the diameter, 2a, was approximately 3 times the effective thickness of the overburden (equation 8). The models suggest that a ratio of 0.8 to 1.5 might be more nearly correct for the right-hand side of equation 8 (Chapter V) when body forces are included in the solution.

No concentration of shear stress is found near the surface at the axis of the models, as Pollard and Johnson (1973) predicted, and no field evidence exists for high shear stresses near the central area of the roof of laccoliths. Instead, hoop stresses away from the axis and near the surface dominate the problem and field examples are known (e.g., Corry and others, 1988). At depth, the highest shear stress is found at the periphery of the sill. In model five, the maximum differential shear stress is 10.2 MPa, and it is found at the periphery of the sill.

Initial failure in model five is by radial and tensile fractures at the surface (Plate VI). The concentration of rotated stress trajectories above the periphery of the sill reflects the concentration of shear stress at the periphery and ensures that when failure occurs the laccolith will shear its periphery vertically. The pattern of the stress trajectories for model five is similar to those for models one and two (Plate VI), and doubling the depth of intrusion appears to have had no effect on the type of failure. Since more ductile behavior would normally be expected for deeper intrusions, any further increase in the depth of intrusion would probably result in significant variation in the type of failure; it is likely that an intermediate shape laccolith would result. This possibility has not yet been modeled, but I suggest that punched laccoliths probably only result from relatively shallow intrusions (<2.5 km).

Sill propagation in a brittle regime. Previous authors have noted that fractures have apparently formed in a brittle regime in the vicinity of intrusions that occurred at depths where the country rock should have been ductile. Generally, this has been attributed to pore pressure, which can lower the effective stresses. In reviewing propagation mechanisms for extension fractures, Pollard (1973, p. 249) suggested that "Pore pressure in the immediate vicinity of the intrusion may be significantly affected by heat from the magma or a volatile phase of the magma if this phase can penetrate into host rock pores." The lack of any significant hydrothermal alteration or thermal metamorphism at the margins of laccoliths contradicts his hypothesis, at least as applied to laccolithic intrusions. Instead, based on my models, I propose that at the outer periphery of the sill the body forces are locally relieved by the sill mechanically lifting the overburden. With the boundary conditions imposed for model five, the overburden becomes transitional in material behavior at a depth of approximately 2,100 m. However, because of the lift of the intrusion, the body forces are relieved for some distance in advance of the intrusion. Two integration stations of the thin element adjacent to the end of the sill are in actual extension in model five. By mechanically relieving the body forces, a local brittle regime is created in front of the sill, and the propagation of a brittle extension fracture is facilitated. The lifting of the sill margin is also seen in Christmas-tree model one, the physical models, and in the field as discussed in model six.

Model six

The depth of intrusion is 1,200 m, and the magma driving pressure, P_d, is 25 MPa as in models two and four. Body forces are included.

Since the state of stress underground is not hydrostatic (Engelder and Sbar, 1984), a model was developed to examine the effect of horizontal stresses on the emplacement. Based on results from hydrofracturing carried out in oil wells in sedimentary basins, Handin (personal communication, 1976) suggested that a ratio of horizontal stress to vertical stress on the order of 0.8 might be appropriate. For the boundary conditions used, a ratio of horizontal to vertical stress of 0.8 corresponds to a Poisson's ratio of 0.45. Based on the same tangent moduli, and at the same strain points, the bulk and shear moduli were recalculated as shown on Plate VI. Model six is then the same as model two except for a change in moduli.

The radius of the sill at failure is 800 m, an increase of 267 m, or 33 percent. The resolution of the value of the radius at failure is about one-half an element width (~130 m). The elements in models two and six are 267 m wide, and the increase in radius thus corresponds to the width of one element. This may be fortuitous and could be the result of the method of analysis. Since, for reasons of economy, convergence tests were not run, the increase in radius must be regarded as qualitative.

The differential shear stress at the margin of the sill is 5.2 MPa. The much greater rotation of the stress trajectories in this model, as compared to model two, is a result of the greatly increased Poisson's ratio. If Poisson's ratio was further increased from 0.45 to 0.5 (i.e., incompressible), the orientation of the stress trajectories would be independent of body forces and the stress trajectories would then be identical to model four. What is suggested by models four and six is that the closer the properties of the overburden approach those of an incompressible substance, the further the laccolith can spread at a given depth without roof failure. Thus, diameter of the laccolith may in part be a function of the horizontal stress field. Since the horizontal stresses may be asymmetric, a laccolith forming in such an environment may then also be asymmetric. Pollard and Muller (1976) have done a preliminary field and theoretical analysis, and model six supports their work.

Roof failure in model six is again by radial and vertical tensile fracture at the surface. From the lower differential shear stress at the periphery of the sill, it is unclear whether this model will fail in pure shear at the periphery, or drape fold the overburden as suggested by the striking resemblance of the stress trajectories of model six (Plate VI) to those of the physical model discussed below.

Comparison of physical and theoretical models.
A set of four physical models was constructed of two layers of Indiana Limestone with a layer of Coconino Sandstone sandwiched between. These physical models were suggested by field experience in the Solitario, a large symmetrical laccolith in west Texas where the flanks of the laccolith are distinctively drape folded (Corry, 1972; Corry and others, 1988). One of these models (C-4) is shown in cross section in Figure 41. The block shown at the bottom left in Figure 41 is forced upward along a precut, lubricated surface into the sandwiched layers under a confining pressure of 100 MPa at a strain rate of 10^{-4}. The experimental technique is described in detail in Min (1974) and Friedman and others (1976). By mapping the microfractures, as shown in Figure 42, which form parallel to the maximum compressive stress, it is possible to derive a stress trajectory diagram as shown in Figure

43. A field example of similar deformation is shown in Figure 44.

Min (1974) has shown that, for an elastic model, principal stress trajectories are nearly independent of the applied displacement along the lower boundary. Thus, it is reasonable to compare the stress trajectories of Figure 43 with those of model six (Plate VI), at least on the downthrown side of the models. Note that there is a scale difference of 10^4 to 10^5 between the field example, the physical model, and the theoretical model. Nonetheless, the resemblance is striking. The areas in extension and compression correlate well. The discontinuities in the stress field above the margin of the uplift are also evident in both models, though obviously more pronounced in the physical model, where through-going fractures have developed. Also note that the bottom layer on the downthrown side of the block (Fig. 41) has been mechanically lifted. Similar behavior is apparent in the areas in model five (Plate VI) that have been lifted by the intrusion and thus become brittle owing to the relief of the body forces.

CHRISTMAS-TREE LACCOLITH MODEL

I have shown above that single-level intrusions can develop strong differential shear stresses at their periphery, which would result in shear fracture there. Such peripheral fractures are evidenced by radial faults around punched laccoliths in the field. I have argued that the lack of peripheral faults and the smooth domes characteristic of Christmas-tree laccoliths can be explained by a multiple-level loading in a Christmas-tree arrangement. Such loading will eliminate the large shear stresses at the periphery of the single sill by stress interference between closely spaced sills. The idea is shown schematically in Figure 8. A field example of a Christmas-tree laccolith is shown in Figure 10, and the theoretical model is shown in Figure 40. The model is again assumed to be axisymmetric for analytical convenience.

Model one

Sills are assumed to intrude at 1,600 m, 1,800 m, 2,000 m, and 2,200 m. A sill separation of 200 m corresponds approximately to the largest separation observed in the field. The sill separation must be much less than the radius of the lacolith if stress interference between the sills is to occur. The magma driving pressure, P_d, is the same in all the sills, 50 MPa. The moduli and Poisson's ratio are shown on Plate VI and are the same as for punched model one.

Loading is such that all four axial elements are loaded initially, after which the sill at 2,200 m grows the fastest. The configuration shown on Plate VI is just prior to roof failure, and only one small radial vertical crack has formed at the surface near the axis. As growth continued, this crack would correspond to the crestal graben illustrated in Figures 8 and 10.

The radius of the deepest sill at failure is 1,400 m as compared with a radius of about 1 km for a punched laccolith at the same depth of intrusion. Thus, the model suggests that multiple-

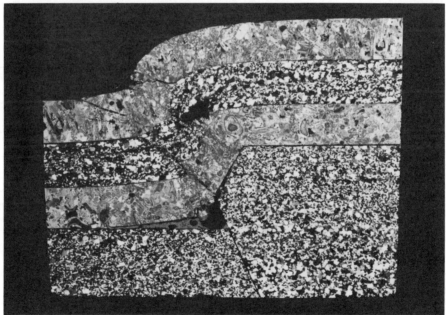

Figure 41. Experimental force fold under conditions simulating the edge of a laccolith. Model C-4. Above, oblique view of model C-4 after deformation. Below, photomicrograph of longitudinal cut through model C-4. Photomicrograph by M. Friedman, 3 × magnification.

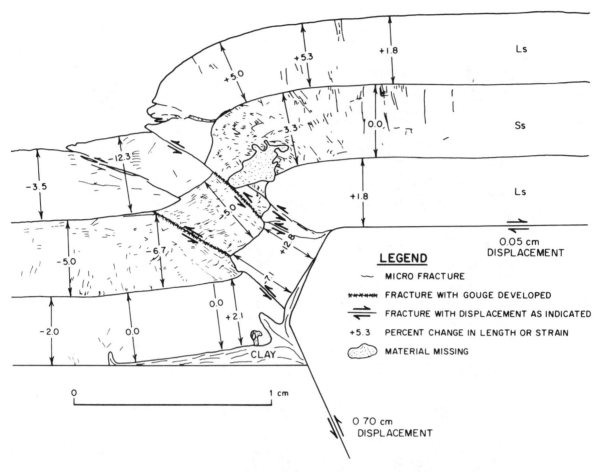

Figure 42. Microfracture map of the experimental force fold shown in Figure 41. Strains are calculated from the relation $\Delta 1/1$. The clay originally filled the notch before deformation was initiated. Compare the intrusion of the clay under the uplifted block during deformation to the magma intrusion under a similar block at Wax Factory laccolith near Terlingua, Texas, shown in Figure 44.

level intrusions will grow to a greater diameter, all else being equal.

The stress trajectories are not strongly rotated at the periphery of any sill, indicating that significant discontinuities in the stress field are not associated with the peripheries of the sills. A complex shear stress pattern exists around the upper sills, but I have not found a convenient and clear way to display it.

I proposed previously that the reason shear failure does not occur is because of interfering plastic zones developing between the sills; this is clearly shown on Christmas-tree model one on Plate VI. I have also assumed that the shallower sills interact with the Earth's surface before the deeper sills do. The stress trajectories converge about the end of the second sill, indicating that the shallow sills do interact with the surface first.

Beyond the end of the deepest sill the body forces have been relieved sufficiently to take the material back into the brittle regime. In contrast to punched model five (Plate VI), the brittle

region extends for 1.2 km beyond the end of the deepest sill, indicating uniform bending of the overburden to a distance >1 km from the end of the intrusion.

It is not straightforward to show from one model that the loading results in formation of the smooth domes observed in the field (Fig. 10), but my assumptions are borne out by the model, and there is no suggestion in the theoretical analysis that shear fractures will develop.

ROOF DEFLECTION

As indicated both by field evidence and the theoretical models, at some point in the loading history of the protolaccolith the deflection of the roof becomes unstable. That is, for an additional load increment, large deflections occur without any linear relation to the load. This nonlinear, inelastic, large-scale deflection of the roof distinguishes laccolithic intrusions from sills.

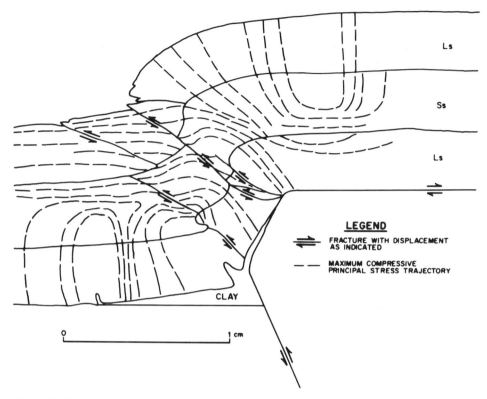

Figure 43. Stress field associated with late-stage deformation of the experimental force fold shown in Fig. 41.

Figure 44. The uplift of the Wax Factory laccolith has rotated the block of Boquillas Formation (Kbo) on the left. Magma intruded beneath the block in the same manner the clay intrudes under the block in the physical model shown in Figures 41, 42, and 43. A hinge zone must have formed above the left side of the uplifted block as in the physical and theoretical models. As uplift continued, the laccolith sheared through the roof rock in the same manner as shown in Figure 17. View is looking south at the northwest margin and the laccolith is about 50 m thick. A more general view can be found in Yates and Thompson (1959, Plate 2).

Unstable behavior is a dynamic process. The theoretical analysis to date has been limited to static analysis. Thus, when unstable behavior occurs, the program becomes numerically unstable. MAGGIE includes dynamic analysis, but the process becomes expensive, and I have had trouble enough with the static analysis. The theoretical investigation of large-scale roof deformation thus remains an unsolved problem.

MAGMA SUPPLY

The laccolith may stop growing before the limiting thickness is reached because the magma supply may be insufficient to form a thick laccolith. As shown in stage one, a typical feeder dike probably contains $\leqslant 0.1$ km^3 of magma. A tenth of a cubic kilometer of magma would be sufficient to raise the roof of a protolaccolith 2 km in diameter just 30 m. Thus, a laccolith may not form, even though roof failure may have occurred, unless an abundant magma supply is available within the freeze time of the sill forming the protolaccolith, typically less than one year.

For a sill 20 km in diameter to raise its roof 100 m requires the addition of more than 30 km^3 of magma to the intrusion. Shaw (1980) estimates the total magma production per year of the entire earth is on the order of 30 km^3. Jordan and others (1986) give an estimate of 300 km^3 per year for the global plate flux. With either estimate, it is evident that the emplacement and growth of a large laccolith consumes a significant portion of the heat budget of the earth during its formation.

CHAPTER VII

SUMMARY AND CONCLUSIONS

Laccoliths are intrusions that spread horizontally high in the lithosphere as thin sills and subsequently deform the roof rock in an unstable, inelastic fashion. An arbitrary, but useful, minimum thickness of 30 m can generally be used to distinguish laccoliths from sills, which are usually 1 to 10 m thick. While sills may rarely be thicker than 30 m, laccoliths described in the literature are always thicker than 30 m. Intrusions with thicknesses greater than 10 m, but less than 30 m, could be classified as protolaccoliths if the roof is mechanically deformed. An example of a proto-laccolith is Trachyte Mesa (Fig. 35) in the Henry Mountains, Utah, which stopped growing just as vertical shear fractures were forming on the periphery.

The known laccoliths of the world have been tabulated in a gazetteer. An extrapolation from the available numbers indicates that there are more than 1,000 laccoliths in North America, and somewhere between 5,000 and 10,000 in the world, unless laccoliths are inexplicably abundant only in North America. Thus, the emplacement and growth of these features, which may range in diameter from 1 km to greater than 100 km, and in thickness from 30 m to 15 km, is one of the major intrusion processes on earth.

A remarkable feature is the rapidity with which laccoliths grow. Calculations of freeze time for the invading sill, and observations of the growth of two laccoliths in Japan, indicate that many laccoliths grow to their full dimensions in less than one year. Almost certainly, only the very largest laccoliths take more than a century to complete their growth. Intrusive activity may continue in the area for considerable time after the laccolith is intruded, but the formation of the laccolith itself is a geologically instantaneous event.

Gravity surveys of a substantial number of felsic laccoliths in the western United States reveal no detectable density contrast between the laccoliths and the country rock. The gravity surveys strongly support Gilbert's (1877) hypothesis that control of the level of intrusion for felsic laccoliths is a function of the density contrast between the rising magma and the weighted mean density of the overburden. Gilbert's hypothesis enables the use of the dislocation model of Weertman (1971a, 1971b) to obtain

reasonable estimates of the magma driving pressure and to establish the boundary conditions for the initiation of the growth of laccoliths. The problem of the control of the emplacement level for mafic sills and laccoliths is by no means as well defined, but Gilbert's model is the only one that fits the available evidence.

The emplacement of laccoliths is dominantly a mechanical process. Evidence for wall rock assimilation or extensive alteration is rarely found in association with laccolithic intrusions. However, small lenses of country rock caught up between sills in Christmas-tree laccoliths may be highly metamorphosed, or partially assimilated.

By establishing the boundary conditions at the level of intrusion, it has been possible to treat the growth of laccoliths as boundary value problems in continuum mechanics. First, however, it was necessary to examine the various types of laccoliths reported in the literature and devise a mechanically reasonable classification scheme. The diverse shapes assumed by laccoliths during the growth phase appear to be a continuous series, with two distinct end members, which I term punched and Christmas-tree laccoliths, as shown in Figure 8.

Punched laccoliths have flat tops, peripheral faults, and steep to vertical sides. Mechanically the intrusion is equivalent to a punch problem with loading of the overburden by a single magma sheet. From a series of finite element models I have demonstrated that such loading results in shear failure at the periphery and produces a punched, cylindrical laccolith. Punched laccoliths are probably formed only by relatively shallow intrusions within the epizone. They commonly have discordant, sheared margins like those of a stock. The presence of a concordant roof ensures that the intrusive is a laccolith, but if the roof is entirely eroded away, geophysical surveys may be required to distinguish stocks from laccoliths.

Christmas-tree laccoliths have no peripheral faults, appear to be rounded domes, and overlying beds are continuous across them. Mechanically they are produced by multiple level intrusions, the deepest of which extends farthest (i.e., a Christmas-tree of sills). Though only one Christmas-tree model has been computed, its stress trajectories indicate that no shear stress concentrations build up at sill peripheries because of the predicted stress interference from nearby sills.

The theoretical models indicate that the diameter of laccoliths is affected by at least the following parameters: (1) the number of sills forming the intrusion; (2) the compressibility of the overburden; (3) the depth of the intrusion; and (4) horizontal stresses. Thus, no simple relation between depth of intrusion and diameter of the laccolith exists. For punched laccoliths a rough correlation between depth and diameter may exist, but no field evidence has been found to support this possibility. Within the limits investigated in the models, the diameter of the laccolith is independent of the magma driving pressure, provided the magma driving pressure exceeds one-half the overburden pressure. If the magma driving pressure is less than one-half the overburden pressure, the roof does not fail, as predicted by Bradley (1965). Thus, most laccoliths form in the epizone. In the mesozone, deformation style changes and lopoliths form. The transition from laccolith to lopolith is continuous as a function of depth, and they are both laccolithic intrusions. The models also suggest that a laccolith will not form if the overburden is incompressible or the horizontal stresses equal or exceed the vertical lithostatic stress.

The theoretical analysis, physical models, and one field example all indicate that body forces in front of the advancing sill will be relieved by mechanically lifting the overburden. The local response of the country rock to the invading intrusion will thus be brittle. Since cracks propagate easier in a brittle medium, the advance of the sill is facilitated.

Nearby large laccoliths will greatly modify the local stress field and probably the form of subsequent intrusions (Fig. 9). It is reasonable to assume that the first laccoliths to form in a group will have the largest magma supply available to them and thus will be the largest laccoliths of the group. If an existing laccolith has not frozen by the time the next intrusion sequence begins, the preexisting laccolith will be fed by any new dikes which come up beneath its floor and continue to grow. Only those dikes that rise outside the existing laccoliths, or cut frozen laccoliths, can form new laccoliths. Late dikes rising around a preexisting intrusion tend to be radially distributed around the central intrusion due to the stress field induced by the initial intrusion. If these dikes feed laccoliths, the resultant laccoliths will also be radially distributed about the central laccolith, as in the Henry Mountains, Utah. The shape of these latter intrusions may be grossly modified by the often extreme deformation accompanying the formation of the earlier laccoliths. Daly (1914, 1933) introduced the term chonolith to describe such irregular intrusions.

Laccoliths intruded into weak layers of water-saturated clays or shales may shed their roofs by gravitationally driven slumping as they grow upward. Such a process would allow them to become indefinitely thick or even extrude onto the surface.

Research on a geologic problem is seldom complete, and this paper is probably best regarded as a centennial progress report on research since the pioneering work of Gilbert (1877).

Only a few of the laccoliths in the world appear to be described in the geological literature, and many critical relationships in the intrusion process remain to be discovered. Quantitative field work—and a good horse—will be prime requisites for the discovery of these relationships.

APPENDIX A

DEFINITIONS OF DESCRIPTIVE TERMS FOR LACCOLITHIC AND ASSOCIATED INTRUSIONS

Definitions in italics are those given by Bates and Jackson (1980). The North American Commission on Stratigraphic Nomenclature (NACSN) is presently (mid-1987) examining a proposal to ". . .eliminate the use of form terms in formal nomenclature; specifically that the form terms batholith, intrusion, pluton, body, stock, dike, sill, diapir and plug be used only as informal terms." according to Susan A. Longacre in a letter to Geotimes (May 1987, p. 2). Informal usage of these terms is followed here, in accordance with the NACSN proposal.

akmolith: *An igneous intrusion along a zone of décollement, with or without tonguelike extensions into the overlying rock.* Coined by Erdmannsdörfer (1924) to describe intrusions in the southern Andes. I have not been able to locate these features. The word has not been generally used and the proposed mechanics seem dubious in light of current knowledge. I favor abandoning the term.

batholith: An informal term for *A large, generally discordant plutonic mass that has more than 40 sq mi (100 km²) of surface exposure and no known floor. Its formation is believed by most investigators to involve magmatic processes.* Laccolithic intrusions may grow to the dimensions of batholiths.

bysmalith: *A roughly vertical cylindrical igneous intrusion. It has been interpreted as a type of laccolith.* Coined by Idding (1898). Paige (1913) and Darton and Paige (1925) purported to explain the formation of bysmaliths by a progressive increase in magma viscosity during the intrusion of the laccolith. Pollard and Johnson (1973) have shown, using mechanical arguments, that Paige's model is inaccurate. Because of the linkage of the term with an invalid mechanical model, I favor abandoning the term. I have used the term "punched laccolith" to describe this type of intrusion in order to associate the name with the responsible mechanical process.

cactolith: *An irregular intrusive igneous body of obscurely cactus-like form, more or less confined to a horizontal zone and appearing to consist of irregularly related and possibly distorted branching and anastomosing dikes that fed a laccolith.* Term introduced by Hunt and others (1953, p. 151): *"a quasi-horizontal chonolith composed of anastomosing ductoliths whose distal ends curl like a harpolith, thin like a sphenolith, or bulge discordantly like an akmolith or ethmolith."* Except for cocktail conversation, the term has little utility and may be Hunt's gentle chiding of geologists for their propensity for inventing descriptive names where possibly existing terms would do. For example, while not so colorful, the term chonolith would serve adequately to describe this type of intrusion.

chonolith: *An igneous intrusion whose form is so irregular that it cannot be classified as a laccolith, dike, sill, or other recognized body.* Introduced by Daly (1914) in recognition of the complex form sometimes assumed by igneous bodies that defy classification on the basis of known structural relations. The term is preferable to "pluton" when describing structural relations of intrusions, as it implies the investigator has examined the structural relations, whereas pluton does not.

Christmas tree: *A term applied to a laccolith or volcanic neck in which sill-like intrusive structures taper away from a central intrusive mass, the whole structure resembling the outline of a cedar tree in cross section.* Bates and Jackson (1980) call this a cedar-tree laccolith and it is also synonymous with a compound laccolith. I prefer the term Christmas tree because of the more evocative imagery.

dike: An informal term for *A tabular igneous intrusion that cuts across the bedding or foliation of the country rock.* It may be difficult in the field to distinguish between a dike intruding at a low angle and a sill that is crosscutting bedding. The problem may be further complicated if there has been postemplacement tectonics in the area.

ductolith: *A more or less horizontal igneous intrusion that resembles a tear drop in cross section.* The term was introduced by Griggs in Hurlburt and Griggs (1939) for intrusions in the Highwood Mountains, Montana. He applied the term to horizontal plugs of teardrop cross section, or headed dikes. The term is seldom used, and probably should be abandoned.

ethmolith: *A discordant pluton that is funnel-like in cross section.* According to Daly (1933, p. 102) the term was introduced in 1903 to describe funnel-shaped intrusions. The term predates Grout's (1918) definition of lopoliths, which, according to the diagrams reproduced in Daly (1933), an ethmolith most resembles. Modern usage calls them simply funnel intrusions and the term should be abandoned.

harpolith: *(a) A large, sickle-shaped igneous intrusion that was injected into previously deformed strata and was subsequently deformed with the host rock by horizontal stretching or orogenic forces. (b) Essentially a* phacolith *with a vertical axis.* These intrusions also resemble trap-door laccoliths. The term is not in general use and the complex structural relations would probably result in such an intrusion being called a lopolith today.

laccolith: *A concordant igneous intrusion with a known or assumed flat floor and a postulated dikelike feeder commonly thought to be beneath its thickest point. It is generally plano-convex in form and roughly circular in plan, less than five miles in diameter, and from a few feet to several hundred feet in thickness.* Laccoliths commonly have much greater dimensions than those given and they may have discordant floors and margins. Laccoliths are distinguished from sills by nonlinear, inelastic, large-scale deflection of the roof. I have used the arbitrary distinction between a sill and a laccolith as a thickness ⩾30 m. No evidence is available to support Billings (1972) classification of a laccolith versus a sill based on a diameter to thickness ratio. The term may also be used in a generic sense for forcible intrusions of any final form that have domed the country rock above or below them and created a chamber.

lopolith: *A large, concordant, typically layered igneous intrusion, of plano-convex or lenticular shape, that is sunken in its central part owing to sagging of the underlying country rock.* Introduced by Grout (1918). The Duluth gabbro is the type example. Lopoliths are a type of laccolithic intrusion (Grout, 1918), and there is no clear-cut distinction between the form of a laccolith and a lopolith.

phacolith: *A minor syntectonic concordant concavo-convex intrusion within folded strata.* The term is infrequently used. The upper layers of a Christmas-tree laccolith may resemble a phacolith at intermediate stages of erosion.

pluton: An informal term for *An igneous* intrusion. *A body of rock formed by metasomatic replacement. . . .The term originally signified only deep-seated or plutonic bodies of granitoid texture.* A much abused term currently in vogue to describe an igneous mass when an author wishes to avoid any discussion of the structural relations of the body. An acceptable, informal term for cases in which the field investigator is not concerned with the structural relations when studying igneous bodies. However, the usage of the term should generally be limited to its original meaning. An epizonal laccolith is not a pluton. If structural relations are unclear, or complex, but forcible injection of the magma is suggested by the field relations, then the body is better termed a chonolith in a structural context.

sill: An informal term for *A tabular igneous intrusion that parallels the planar structure of the surrounding rock.* Sills may also crosscut bedding but at the time of intrusion they are often injected along a plane sensibly parallel to the earth's surface.

sphenolith: *A wedgelike igneous intrusion, partly concordant and partly discordant.* The term is infrequently used and its utility is limited.

stock: An informal term used to describe *An igneous intrusion that is less than 40 sq mi (100 sq km) in surface exposure, is usually but not always discordant, and resembles a batholith except in size.* Laccoliths with discordant margins are frequently mistaken for stocks. The presence of concordant roof pendants from shallow depths is evidence that the intrusion is a laccolith rather than a stock. Felsic stocks normally have a distinct gravity anomaly, whereas felsic laccoliths do not.

trap-door fault: *A curved fault bounding a block that is hinged along one edge; it is an* intrusion displacement *structure in the Little Rocky Mountains of Montana.* More commonly called trap-door laccoliths. Any laccolith that has a tilted roof with one, or more, sides defined by a fault or drape folding, with at least one side folded rather than faulted. A very common type of laccolithic intrusion. Hyndman (personal communication, 1987) suggests that many of these types of laccolith have lateral feeder dikes along the bounding fault. Preexisting structure, zones of weakness, or faults appear to control roof failure in trap-door laccoliths.

APPENDIX B

A GAZETTEER OF LACCOLITHS

COUNTRY State, Group, and Laccolith Name	County or District	Latitude	Longitude	Length km	Width km	Thick- ness km	Depth of Intrusion km or fm	Topographic Map	References and Remarks
ALGERIA									
Djebel Aroudjaoud		36°37′N?	2°22′E?					Cherchel	Glangeaud (1934)
ARGENTINA									
C°Blanco de Zonda		31°30′N	68°50′W						Leveratto (1968)
ANTARTICA									
Dufek		82°36′S	52°30′W						Behrendt and others (1974)

Additional references: Abel and others (1979), Behrendt and others (1979), England and others (1979), Ford and others (1979), Kistler and Ford (1979)

COUNTRY State, Group, and Laccolith Name	County or District	Latitude	Longitude	Length km	Width km	Thick- ness km	Depth of Intrusion km or fm	Topographic Map	References and Remarks
AUSTRALIA									
New South Wales									
Montagne Island ?	Narooma	36°16′S	150°12′E						Jevons and others
Mt. Dromedary	Narooma	36°18′S	150°00′E	14.5					(1911, 1912)
Prospect Mountain		33°49′S	150°54′E	3	1.6	> 0.1	0.2 +		Brown (1930) Wilshire (1967)
Tasmania									
Cygnet		43°09′S	147°04′E						Leaman and Naqui (1967)
Bruny Island		43°17′S	147°18′E						Stephens (1902)
D'Entrecasteaux Channel		43°17′S	147°15′E						
Forestier Penninsula		42°57′S	147°55′E						
Mt. Connection		?	?						
Mt. Dromedary		42°43′S	147°07′E						
Mt. Field East		42°39′S	146°39′E						
Mt. Field West		42°39′S	146°32′E						
Mt. Seymour		42°22′S	147°28′E						
Mt. Tooms		42°13′S	147°53′E						
Quamby's Bluff		41°40′S	146°42′E						
Tasman's Penninsula		43°05′S	147°50′E						
Mt. Olympus		42°03′S	146°07′E						
Western Australia									
Panton sill		17°46′S	127°50′E	10	3	1.5			Hamlyn (1980)
Wellington Range		26°17′S	121°50′E						
BRAZIL									
Tambau		47°05′N	22°42′E						Davino (1980)
CANADA									
British Columbia									
Cariboo-Bell	Cariboo	52°30′N	121°25′W	6		2 to 3			Hodgson and others (1976)
Ice River	Yoho National Park	51°20′N	123°30′W						Allan (1914)
Nimpkish batholith		50°02′N	126°25′W						Gunning (1932)
Labrador									
Kiglapait		57°00′N	61°28′W	32.4	26.7	9.5			Morse (1969) Stephenson and Thomas (1979)
Michikamau		54°30′N	64°00′	> 60		> 15			Emslie (1970)

COUNTRY State, Group, and Laccolith Name	County or District	Latitude	Longitude	Length km	Width km	Thick-ness km	Depth of Intrusion km or fm	Topographic Map	References and Remarks
									Lopolith
Manitoba									
Mikanagan Lake		54°55′N?	101°35′W?						Bateman and Harrison (1944)
Newfoundland									
Trout River group									
North Arm Mountain		49°17′N	58°00′W	14	9.2				Ingerson (1935) Buddington and Hess (1937), Ingerson (1937) Ultramafic laccoliths
Table Mountain		49°25′N	58°00′W	12.4	8.4				
Northwest Territories									
Coppermine River		?	?						Tedlie (1960)
Great Slave Lake group									
Meridian Lake	Mackenzie	62°37′N	109°30′W						Hoffman (1968) Badham and Nash (1978)
Regina Bay	Mackenzie	62°27′N	110°18′W						
Additional references: Gandhi and Prasad (1980, 1982), Badham (1981)									
Dubawnt group									
Angikuni Lake	Keewatin	62°25′N	99°40′W						Blake (1980)
Baker Lake	Keewatin	63°45′N	95°30′W						
Approximately 34 laccoliths are in the vicinity of Baker Lake. An additional 10 laccoliths occur in the vicinity of Angikuni Lake.									
Ontario									
Bull Lake	Algoma	~46°N	~81°W						James (1984) James and Born (1985)
Quebec									
Clericy ?	Clericy	48°20′N	78°50′W	11	1.6				Cooke (1930) Cooke and others (1931)
Lake Dufault	Dufresnoy	48°20′N	79°00′W	9.6	8				
Several other bodies in this area may be laccoliths but structural relations are not given.									
Lake Chibougamau		49°50′N	74°20′W	>45					Fairbault and others (1911), Daly (1914)
Monteregian Hills?		45°33′N	73°09′W						Clark (1972)
A group of ten hills may be laccoliths but details are not available.									
Saskatchewan									
Carswell		58°27′N	109°30′W	39	39				Currie (1969)
CHINA									
Fuhsing	Taiwan	24°45′N	121°20′E	1		0.3			Yang and Lee (1978)
CZECHOSLOVAKIA									
Moldau River area		49°52′N	14°21′E	10	5				Kettner (1914)
EGYPT									
Bula		26°41′N	33°32′E	6					El-Gaby and Habib (1978)

COUNTRY State, Group, and Laccolith Name	County or District	Latitude	Longitude	Length km	Width km	Thickness km	Depth of Intrusion km or fm	Topographic Map	References and Remarks
Nile Valley group									
Gabal El Fahdi		32°33′N	26°48′E	~5	~4	>.13			El Tahlawi (1974)
Gabal Gebeil		30°31′N	27°24′E	~7	~5	>.15			
Kolet Abu Gilbana		31°53′N	26°33′E	~6	~4	>.12			
Kolet El Fartella		31°38′N	27°06′E	~7		>.16			
FINLAND									
Ahvenisto		61°30′N	26°30′E						Laurén (1970)
Laitila		60°52′N	21°40′E						
Vehmaa		60°41′N	21°40′E						
Viipuri		60°50′N	27°00′E						
Savonlinna		62°30′N?	28°00′E?	18	6				Parkkinen (1975)
FRANCE									
La Faurie		45°15′N	1°27′E						Santallier and Floc'h (1979)
Longue Island		48°15′N	4°35′W						Babin and others (1968)
Margeride		45°00′N	3°10′E						Couturie and others (1979) Autran and others (1979)
Mayet de Montagne		46°05′N	3°40′E						Didier and Peyrel (1980)
Mont Lozere		44°21′N	3°35′E						Fernandez (1977)
Mont Pilat	Veranne	45°23′N	4°35′E						Chenevoy (1967)
Pic du Midi d'Ossau		42°51′N	0°26′W						Bixel (1973)
Sacharoide		45°42′N	3°01′E						Peyronnet (1964)
Massif Central group									
Affleurement de Lascaux	Limousin	45°40′N	0°55′E						Menot and Piboule (1977), Vauchelle and Lemeyre (1983)
Massif de La Boissone	Limousin	45°40′N	0°55′E						
Massif de Champagnac-la-Rivière	Limousin	45°42′N	0°55′E						
Massif del Coquille	Limousin	45°34′N	0°59′E					Chalus XIX-32	
Massif de Crezeunet	Limousin	45°03′N?	0°30′E?						
Massif de Cussac	Limousin	45°42′N	0°51′E						
Massif Gueret	Limousin	46°N?	2°E?						
Massif de Merly	Limousin	45°45′N	0°49′E	1.5	0.5				
Massif du Puy de Lageyrat	Limousin	45°40′N	0°56′E						
Massif del Rougerie	Limousin	45°22′N	0°50′E					Chalus XIX-32	
Massif de Saint Bazile	Limousin	45°44′N	0°49′E						
GERMANY									
Bavaria									
Eulengebirge		50°35′N?	16°40′E?						Schwan (1968, 1976)
Frankenberg		50°48′N?	13°10′E?						
Grafenthaler?		50°30′N?	11°23′E?						
Hirschberg		50°20′N?	11°55′E?						
Munchberg		50°05′N?	11°45′E?						
Wildenfels		50°33′N?	12°40′E?						
Uchte		52°20′N	8°50′E						Nodop (1971)

COUNTRY State, Group, and Laccolith Name	County or District	Latitude	Longitude	Length km	Width km	Thickness km	Depth of Intrusion km or fm	Topographic Map	References and Remarks
Wurttemberg									
Steinheim Basin		48°58′N?	9°16′E?						Adam (1979)
GREENLAND									
Augpilagtoo		60°20′N	44°30′W						Ussing (1911)
Frederiksdal		60°N	45°W						Escher (1966)
Grahs Fjelde		61°N	43°W						Wager and Brown (1967)
E. of Patussoq		60°50′N	43°45′W						Bridgwater and others
Prins Christian Sund		60°10′N	43°15′W						(1974)
Skaergaard		60°15′N	31°15′W	~5		1			Lopoliths
Sydproven		60°20′N	45°15′W						
GUYANA									
Mt. Roraima		5°14′N	60°44′W						Harrison (1909)
Tumatumari-Kopinang sill		5°N	59°30′W			0.15-0.55			Hawkes (1966)
HUNGARY									
Csodi Mountain		46°15′N?	18°05′E?						Buda (1966)
Erdobenye		48°20′N?	21°35′E?						Kulcsar and Barta (1971)
Hajnac Mountain		48°00′N?	19°48′E?						Poka and Simo (1966)
Matra Mountains		47°55′N?	20°00′E?						Pesty (1966)
ICELAND									
Sandfell		64°55′N	13°56′W			0.6	0.55		Hawkes and Hawkes (1933)
INDIA									
Bihar									
Neropahar		24°28′N	85°38′E	1.5	.37	>.15			Chattopadhyay and Saha (1974)
Gujarat									
Barda Hills	Saurashtra	21°50′N	69°45′E						Dave (1972)
Girhar Hills	Saurashtra	21°34′N	70°29′E						Rao (1967)
South Andaman Island									
Chaulduari		11°50′N?	92°40′E?						Raina (1974)
Goa									
San Pedro		15°27′N?	73°55′E?						Wagle and Almeida (1974)
Kutch									
Chanduva Hills ?		23°15′N	69°30′E						Blake (1898)

32 laccoliths are reported to exist in the state of Kutch.

COUNTRY State, Group, and Laccolith Name	County or District	Latitude	Longitude	Length km	Width km	Thickness km	Depth of Intrusion km or fm	Topographic Map	References and Remarks
IRELAND									
Donegal		54°55′N	8°00′W						Kennan (1979)
Leinster		52°37′N	6°47′W						
ITALY									
Basilicata									
Euggannean Hills		45°19′N	11°40′E					Padova F. 50 Rovigo F. 64	Cornu (1906) Stark (1907, 1912) Lachman (1909)
Sinni Valley group									
F. del Monaco		40°06′N	16°04′E					Sant′ Arcangelo	Viola (1892)
M. Alpi ?		40°07′N	15°59′E					Sant′ Arcangelo	Daly (1914)
Manca di Sopra		40°04′N	16°06′E			0.3		Sant′ Arcangelo	Mafic intrusions.
Sardegna									
Calabrian Massif		38°N?	16°E?						Schenk and Lattard (1981)
Campidano Graben		39°30′N?	8°47′E?						Pecorini (1966)
Piano Lasina		?	?	2.5					
Cecina Valley group									
Montecatini		43°22′N?	10°45′E?						Mazzanti and others (1963), Squarci and Taffi (1963)
Orciatico		43°25′N?	10°41′E?						
JAMAICA									
Port Royal	Surrey	18°03′N?	76°43′E?					Kingston	Trechmann (1942)
JAPAN									
Hokkaido									
Hidaka Mountains		42°10′N?	143°00′E?						Nagasaki (1966)
Usu-san group									
New Mountain (1910)		42°33′N	140°50′E	2.7	0.6	0.1 presently 0.155 at maximum			Omori (1911a, 1911b) Bailey (1912)
Syowa New Mountain (1944-1945)		42°32′N	140°51′E?	1.0	0.8	0.20 maximum 0.17 presently			Oinouye (1917) Minakami and others (1951), Yagi (1953)

The growth of both these laccoliths was observed and recorded.

COUNTRY State, Group, and Laccolith Name	County or District	Latitude	Longitude	Length km	Width km	Thickness km	Depth of Intrusion km or fm	Topographic Map	References and Remarks
LIBYA									
Kaf Mantrus		32°10′N?	13°E?						Piccoli and Spadea (1964)
MEXICO									
Chihuahua									
Cerro de Cristo Rey		31°47′N	106°33′W	3					Lovejoy (1976)
Cerro de la Mina		31°45′N	106°33′W						

COUNTRY State, Group, and Laccolith Name	County or District	Latitude	Longitude	Length km	Width km	Thickness km	Depth of Intrusion km or fm	Topographic Map	References and Remarks
Coahuila									
Big Bend group (continues north into Texas)									
Cerro Agua Chile	Acuna	29°15′N	102°31′W						Daugherty
Cerro del Colorado?	Acuna	29°15′N	102°09′W						(1963a, 1963b)
Cerro del la Hormiga	Acuna	29°24′N	102°38′W						
Cerro del Veinte	Acuna	29°23′N	102°43′W						
La Cueva	Acuna	29°23′N	103°38′W	6.4	4.8	1.0			
Pico Etero	Acuna	29°23′N	102°35′W						
Nuevo Leon									
East		26°07′N	99°56′W?	6.4	4.8				McKnight
West		26°07′N	99°58′W?	4.8	4.0				(1963a, 1963b)
San Luis Potosi									
Sierra del Fraile		25°55′N	100°35′W			1.5			Spurr and others (1912)
Tamaulipas									
San Carlos Mountains group									
Cerro Diente		24°32′N	98°49′W						Watson (1937)
Cerro Jatero		24°38′N	98°47′W						
Cerro Sacramento ?		24°42′N	98°43′W			> 0.2			
Marmolejo Ranch		24°38′N	98°52′W						
Sierra de Patado		24°38′N	98°32′W						
Sierra de San Jose		24°42′N	98°51′W						
San Miguel		24°44′N	98°58′W						
Tiray		24°39′N	98°52′W						
Zacatecas									
Diguiyu		17°34′N	90°50′W						Ochoterena (1981)
Sierra de Mazapil group									
La Caja		24°33′N?	101°38′W?						Burckhardt (1906)
Las Parroquias		24°35′N?	101°37′W?						
Puerto del los Novillos		24°37′N?	101°38′W?						
Santa Rosa (Cerro Colorado)		24°34′N?	101°31′W?						
NEW ZEALAND									
Otago		46°10′S	170°40′E						Kingma (1974)
Information is lacking on number or precise locations of reported laccoliths.									
Riwaka complex	Northwest Nelson	41°05′S	173°00′E	> 45					Bates (1980)
Cook Island group									
Aitutaki		19°55′S	159°50′W						Summerhayes (1968)
Mangaia		19°55′S	157°55′W						
Manuae		19°15′S	159°W						
Mitario and Mauke		20°S	157°30′W						
Raratonga		21°15′S	159°50′W						
Takutea, Atiu, and Mitiaro		20°S	158°W						
NORWAY									
Bjerkreim		59°36′N	6°08′E						Brögger (1895)
Gjevlekollen ?		59°53′N	10°11′E						Krause and Pape (1977)
Sogndal		58°20′N	6°18′E						Duchesne (1978)
Storstenfjeld ?		59°52′N?	7°43′E?						

COUNTRY State, Group, and Laccolith Name	County or District	Latitude	Longitude	Length km	Width km	Thickness km	Depth of Intrusion km or fm	Topographic Map	References and Remarks
Sulitjelma gabbro		67°10'N	15°20'E	13	10				Mason (1971) 30 m contact aureole
OCEANIA									
New Caledonia (French)									
Pic Du Rocher		22°00'S	166°30'E						Deneufbourg (1969)
South Pacific									
A		19°32'S	120°34'W						Lonsdale (1983)
B		19°32'S	120°57'W						
C		19°33'S	121°16'W						
D		19°33'S	121°25'W						
E		19°32'S	121°30'W	5		.3			
F		19°33'S	121°37'W						
G		19°33'S	121°41'W						
POLAND									
Sudeten Mountains group									
Klodzko Zloty		50°00'N?	15°00'E?						Cwojdzinski (1975)
PORTUGAL									
Serra de Monchique		37°19'N	8°33'W						Rock (1978)
ROMANIA									
Haghita Mountains group									
Budös		46°16'N?	25°45'E?						Rusu (1975)
Cucu		46°13'N?	25°47'E?						
Maciu		46°20'N?	25°43'E?						
SAUDI ARABIA									
Wadi Hayyan and Wadi Qabqab		27°12'N	36°28'E	24	10				Hirokawa (1970) Takahashi (1970) lopolith
SOUTH AFRICA									
Bushveld group									
W. of Balmoral	Pretoria	25°30'S	28°50'E						Molengraff (1901, 1904)
Bushveld complex	Pretoria	25°30'S?	27°30'E?	>450	>270	>8			Brouwer (1910)
E. of Heidelberg ?	Heidelberg	26°40'S	29°10'E						Daly (1914), Hall and Molengraff (1925) du Toit (1954) McBirney (1984)
Cape Province group									
Drooge Fontein ?		30°35'S	26°51'E	2.4	2.4				du Toit (1920)
Jackals Kop		31°21'S?	26°58'E?						
Mt. Arthur ?		30°40'S	27°09'E	11.2	6.4				
Stafelberg's Vley (Limoen Fontein?)		31°29'S?	26°50'E?	9.6	5.6				

COUNTRY State, Group, and Laccolith Name	County or District	Latitude	Longitude	Length km	Width km	Thickness km	Depth of Intrusion km or fm	Topographic Map	References and Remarks
Damaraland group									
Kampaneno or Tjirundu Mts.		?	?						Gevers and Frommurze (1929), Gevers (1963) A large number of intrusive granite domes that are probably lopoliths or laccoliths exist in this area but no details are available.
Lievenberg dome ?		22°12′S?	16°15′E?						
Marble dome		?	?						
Okandura dome ?		22°00′S?	16°12′E?						
Ombujomenge dome?		22°09′S?	16°05′E?						
Otjua dome ?		22°10′S?	16°10′E?						
Waldau Ridge dome ?		22°00′S?	16°40′E?						
Orange Free State									
Trompsberg		30°01′S	25°46′E	70					Buchmann (1960)
SUDAN									
Darfur dome		13°N	24°E	700					Bermingham and others (1983)
SWEDEN									
Bohus		58°25′N	13°E?						Lind (1967)
Galleuaur		66°20′N?	18°13′E?			3.5-4.5			Enmark and Nisca (1982)
Ångermanland									
Nordringrå		62°50′N?	18°20′E?						Högbom (1909)
Ragunda		63°10′N?	16°25′E?						
SWITZERLAND									
Aletsch		46°28′N	8°00′E	30					Baltzer (1904)
Gasteren		?	?	13	8				
Gottard		?	?	72.5	>4.5	>1.5			
TANZANIA									
Mara		1°36′S	34°42′E	45					Darracott (1974)
UNITED KINGDOM									
Dartmoor		51°25′N	4°00′W						Osman (1924), Brammall and Harwood (1932)
Scilly Isles		49°55′N	6°20′W						Osman (1928)
Channel Islands group									
Guernsey		49°28′N	2°31′W						Briden and others (1982)
Jersey		49°10′N	2°06′W						
Scotland									
Arran	Buteshire	55°39′N	5°15′W	12.8	12.8			Knapdale and Firth of Clyde	Bailey (1926) Tyrrell (1928)
Cnoc-na-scroine	Sutherland	58°04′N	4°57′W	>6	4	>0.04		Loch Inver and Loch Assynt	Shand (1910) Read and others (1926)
Loch Ailsh	Sutherland	58°04′N	4°57′W					Loch Inver and Loch Assynt	Parsons (1965)
Isle of Mull	Argyllshire	56°20′N	5°45′W					Sound of Mull	Geikie (1888)

COUNTRY State, Group, and Laccolith Name	County or District	Latitude	Longitude	Length km	Width km	Thickness km	Depth of Intrusion km or fm	Topographic Map	References and Remarks
Kilsyth-Croy	Stirlingshire	56°00'N	4°05'W	> 5		0.1		Glasgow	Tyrrell (1909) Daly (1914)
Loch Borrolan complex		58°11'N?	5°06'W?						Wooley (1970)
Isle of Skye group									
Allt ant-Sithean	Inverness	57°17'N	6°09'W					Portree	Harker (1904)
An Sgùman	Inverness	57°11'N	6°15'W					Rhum, Port of Skye	Some feeder dikes are
Beinn na Caillich ?	Inverness	57°13'N	5°59'W					Portree	exposed. Some laccoliths
Beinn an Dubhaich ?	Inverness	57°12'N	5°58'W	4.4	1.2			South Skye, Arissag	are mafic or ultramafic
Broadford	Inverness	57°16'N	5°55'W					Portree	in composition. Others
Cuillins	Inverness	57°14'N	6°14'W	13.6	9.6	1		Portree	are felsic.
Gars-bheinn	Inverness	57°11'N	6°13'W					Rhum, Port of Skye	
Isle of Soay- An Dubh-sgeire	Inverness	57°09'N	6°13'W					Rhum, Port of Skye	
Red Hills	Inverness	57°17'N	6°06'W			> 0.5		Portree	
Sgùrr Dubh	Inverness	57°12'N	6°14'W			0.46		Rhum, Port of Skye	
Sgùrr Dubh Bheag	Inverness	57°13'N	6°11'W					Rhum, Port of Skye	
Sgùrr na Banachdich	Inverness	57°13'N	6°15'W			0.08		Rhum, Port of Skye	
Wales									
Corndon group									
Corndon Mountain	Montgomery	52°34'N	3°02'W					Ludlow	Watts (1886)
Linley	Montgomery	52°32'N	2°57'W					Ludlow	Lapworth and Watts (1894)
Wilmington Pitcholds	Montgomery	52°32'N	3°00'W					Ludlow	

Other laccoliths are present but no details are available.

UNITED STATES OF AMERICA

Alaska

Amphitheater Mountains		63°07'N	145°45'W					Mt. Hayes 1° × 2°	Rose (1966)

Arizona

Chiricahua Mountains?	Cochise	31°50'N	109°18'W					Chiricahua Peak 15'	Plouff (1958), Drewes and Williams (1973) Newell (1982)
Preston Mesa	Coconino	36°22'N	111°13'W						Breed (1971)
Springerville- Show-Low	Apache and Navajo	34°10'N?	109°45'W?					McNary 15'	Aubele and Crumple (1983)

Carrizo Mountains group

Beclabito dome	San Juan	36°49'N	109°01'W					Pastora Peak and Rattlesnake 15'	Emery (1916) Strobell (1956a, 1956b)
W of Beclabito dome	Apache	36°48'N	109°05'W					Pastora Peak 15'	
Black Rock Point	Apache	36°51'N	109°16'W					Pastora Peak and Toh Atin Mesa 15'	
Carrizo Mountains	San Juan and Apache	36°50'N	109°10'W					Pastora Peak 15'	
Chezhindeza Mesa	Apache	36°49'N	109°01'W			1 ?		Pastora Peak 15'	
North Mesa	Apache	36°52'N	109°09'W					Pastora Peak 15'	
N of Walker Peak	Apache	36°39'N	109°08'W					Tall Mesa 7½'	
Poko or Whirling Mt	Apache	36°44'N	109°12'W					Redrock Valley 15'	
W of Whirling Mt.	Apache	36°44'N	109°14'W					Redrock Valley 15'	

San Francisco Mountains group

Elden Mountain	Coconino	35°15'N	111°38'W	6.4				Flagstaff West 7½'	Robinson (1913), Mintz
Slate Mountain	Coconino	35°30'N	111°51'W					Kendrick Peak 7½'	(1942), Verbeek (1971),
White Horse Hills or Marble Mountain	Coconino	35°25'N	111°40'W			1.2 ?	Redwall ls	White Horse Hills 7½'	Verbeek (1972)

Elden and Slate Mountains reached the surface during emplacement

COUNTRY State, Group, and Laccolith Name	County or District	Latitude	Longitude	Length km	Width km	Thick- ness km	Depth of Intrusion km or fm	Topographic Map	References and Remarks
Santa Rita Mountains group									
Grosvenor Hills	Santa Cruz	31°32'N	110°55'W					Mt. Wrightson 15'	Drewes (1972a, 1972b)
Tejano Springs	Santa Cruz	31°33'N	110°55'W					Mt. Wrightson 15'	
		At least four other unnamed laccoliths exist in the Grosvenor Hills.							
Arkansas									
Magnet Cove	Hot Spring	34°27'N	92°50'W					Malvern 7½	Washington (1901)
California									
Emery Knoll		33°02'N	118°23'W	15		> .45		Long Beach 1°×2°	Junger and Sylvester (1979)
Sutter or Marysville Buttes	Sutter	39°12'N	121°50'W					Sutter Buttes 7½	Williams (1929)
Death Valley group									
Cima dome	San Bernadino	35°16'N	115°35'W	16	16	> 0.3		Mescal Range, Kelso 15'	Ferguson and others (1954), Hewett (1956)
Kingston Peak	San Bernadino	35°45'N	115°57'W	16	16			Kingston Peak 15'	
Leiser Ray Mine	San Bernadino	35°02'N	115°01'W					Lanfair Valley 15'	Johnson (1957)
Little Chief stock?- Hanaupah Canyon	Inyo	36°10'N	117°04'W	9.6	5	1.8		Telescope Peak 15'	McDowell (1974) Hunt and Mabey (1966)
Manly Peak	Inyo	35°55'N	117°09'W					Manly Peak 15'	
New York Mountain	San Bernadino	35°15'N	115°21'W					Ivanpah, Mid Hills 15'	
Skidoo	Inyo	36°26'N	117°09'W	> 19	8	1.1	Kingston Pk?	Emigrant Canyon 15'	
N. of Spring Canyon	Inyo	35°57'N	117°01'W					Manly Peak 15'	
Willow Spring	Inyo	35°54'N	117°04'W					Manly Peak 15'	
Colorado									
Chimney Creek dome	Routt	40°36'N	107°02'W			> 0.06	> 0.55	Wolf Mtn. 7½'	Bass and others
Wolf Creek dome	Routt	40°33'N	107°05'W					Wolf Mtn. 7½'	(1955)
Green Mountain	Summit	39°52'N	106°22'W					Mt. Powell 15'	Heaton (1939)
Two Buttes	Prowers	37°38'N	102°31'W					Two Buttes Reservoir 7½'	Gilbert (1896), Gilbert and Cross (1896) Kreiger (1976), Kreiger and Thorton (1976)
Beaver-Tarryall group									
Alma	Park	39°18'N	106°03'W					Alma 7½'	Singewald (1942)
E. of Alma ?	Park	39°17'N	106°02'W					Alma 7½'	Reconnaissance map,
Boreas Mountain ?	Park	39°23'N	105°56'W					Boreas Pass 7½'	probably many other
Hosier	Park	39°21'N	106°03'W					Alma 7½'	laccoliths in the area.
Little Baldy Mtn. ?	Park	39°20'N	105°56'W					Como 7½'	
Mt. Silverheels- Montgomery Gulch stock?	Park	39°20'N	106°00'W					Alma and Como 7½'	
Palmer Peak	Park	39°19'N	105°59'W					Como 7½'	
Peabodys ?	Park	39°21'N	105°55'W					Como 7½'	
San Juan Mountains group									
The Blowout	Ouray	38°02'N	107°40'W					Ouray 7½'	Holmes (1877)
Canyon Creek	Ouray	38°00'N	107°42'W					Ouray, Ironton 7½'	Cross (1898), Cross
Dexter Creek	Ouray	38°03'N	107°38'W					Ouray 7½'	and Purington (1899)
Deadwood	San Juan	37°49'N	107°43'W	1.6		0.9		Silverton 7½'	Cross and others (1899)
Engineer Mountain	San Juan	37°42'N	107°48'W					Engineer Mt 7½'	Cross and Spencer (1900)
Flat Top	Dolores	37°45'N	107°57'W	> 5		0.46		Hermosa Pk 7½'	Cross and Ransome (1905)
Gray Head	San Miguel	37°58'N	107°57'W	< 5				Grey Head 7½' Telluride and Montrose 15'	Cross and others (1905) Cross and Hole (1910) Crawford (1913)
Howard Fork	San Miguel	37°52'N	107°46'W					Ophir 7½'	Burbank (1930)
Grayrock Peak	San Juan	37°41'N	107°52'W					Engineer Mt 7½'	Eckel (1936, 1937)
Mt. Abrams ?	San Juan and Ouray	37°58'N	107°38'W					Ironton 7½'	Kelley (1945) Larsen and Cross (1956)

COUNTRY State, Group, and Laccolith Name	County or District	Latitude	Longitude	Length km	Width km	Thick- ness km	Depth of Intrusion km or fm	Topographic Map	References and Remarks
Oak and Corbett Cr.	Ouray	38°01′N	107°42′W					Ouray 7½′	Luedke and Burbank (1962)
Silver Mountain	La Plata	37°22′N	108°00′W					La Plata 7½′	Plouff and Pakiser (1972)
Tomichi dome	Gunnison	38°29′N	106°30′W					Bonanza 15′	
Whipple Mountain	San Miguel	38°01′N	107°55′W	< 5				Sams 7½′	

Sleeping Ute Mountain group

Banded	Montezuma	37°12′N	108°47′W	1.6	0.8	0.17	Dakota ss Mancos sh	Mariano Wash E 7½′	Shoemaker and Newman (1953)
Black Mountain stock ?	Montezuma	37°16′N	108°48′W	2.0	2.0			Battle Rock 7½′	Ekren and Hauser (1956, 1965)
N. Black Mountain	Montezuma	37°17′N	108°48′W	3.2	1.2	0.21		Battle Rock 7½′	Case and Joesting
″The Buttes″	Montezuma	37°17′N	108°50′W	1.6		0.15	Dakota ss Mancos sh	Battle Rock 7½′	(1972)
East Horse	Montezuma	37°16′N	108°45′W	2.4	1.6	0.11	Mancos sh	Battle Rock 7½′ and Mud Creek 7½′	
East Sundance	Montezuma	37°14′N	108°47′W			0.11	Dakota ss Mancos sh	Mariano Wash E 7½′	
Flat	Montezuma	37°12′N	108°47′W	2.4	0.53	0.09	Mancos sh	Mariano Wash E 7½′	
Horse Mountain	Montezuma	37°16′N	108°47′W	1.6	0.8	0.34		Battle Rock 7½′	
Irwin	Montezuma	37°14′N	108°44′W			0.11	Dakota ss Mancos sh	Towaoc 7½′	
″The Knees″ stock ?	Montezuma	37°13′N	108°48′W	1.1	0.4			Mariano Wash E 7½′	
Last Spring	Montezuma	37°13′N	108°47′W	1.6		0.23	Dakota ss Mancos sh	Mariano Wash E 7½′	
Mable Mountain	Montezuma	37°18′N	108°47′W			0.46		Battle Rock 7½′	
McElmo dome	Montezuma	37°22′N	108°47′W	32	16	> 0.2	> 1.4	Battle Rock 7½′	
Mushroom	Montezuma	37°14′N	108°49′W	3.6	1.6	0.15	Mancos sh	Mariano Wash E 7½′	
North Ute Peak	Montezuma	37°18′N	108°45′W	> 1.6	0.8	0.18		Battle Rock 7½′	
Pack Trail	Montezuma	37°15′N	108°48′W	2.0	1.6	0.21	Mancos sh	Mariano Wash E 7½′	
Razorback	Montezuma	37°15′N	108°45′W	1.6	0.64	0.26	Mancos sh	Battle Rock 7½′	
Sentinel	Montezuma	37°12′N	108°48′W	3.2	1.2	0.24		Mariano Wash E 7½′	
Sentinel Peak dome?	Montezuma	37°11′N	108°47′W			0.2		Mariano Wash E 7½′	
Sundance	Montezuma	37°14′N	108°47′W			0.11	Dakota ss Mancos sh	Mariano Wash E 7½′	
Three Forks	Montezuma	37°12′N	108°47′W					Mariano Wash E 7½′	
Tongue	Montezuma	37°14′N	108°47′W	0.8		0.06		Mariano Wash E 7½′	
Towaoc dome ?	Montezuma	37°13′N	108°47′W					Mariano Wash E 7½′	
Trap Door	Montezuma	37°14′N	108°47′W			0.11	Dakota ss Mancos sh	Mariano Wash E 7½′	
Ute dome ?	Montezuma	37°15′N	108°49′W	16	16	0.5		Battle Rock 7½′	
Ute Peak stock ?	Montezuma	37°17′N	108°47′W	2.0	2.0	0.76		Battle Rock 7½′	
The ″West Toe″	Montezuma	37°12′N	108°50′W					Mariano Wash E 7½′	
′West Toe′ cluster	Montezuma	37°13′N	108°50′W				Mancos sh	Mariano Wash E 7½′	
Yucca cluster	Montezuma	37°15′N	108°52′W			0.09	Mancos sh	Mariano Wash E 7½′	

Spanish Peaks group

Black Hills	Huerfano	37°39′N	104°59′W	3.22	2.4			Farisita and Black Hills 7½′	Hills (1889a, 1889b) Johnson (1961, 1968)
Gardner Butte or Turkey Creek Butte	Huerfano	37°46′N	105°06′W					Badito Cone 7½′	
Huerfano Butte ?	Huerfano	37°45′N	104°49′W					Huerfano Butte 7½′	
Mt. Mestas or Veta Mountain	Huerfano	37°35′N	105°07′W					La Veta Pass 7½′	
N. Middle, and S. White Peak-Three Buttes?	Huerfano	37°22′N	105°04′W					Cucharas Pass 7½′	
Sheep and Little Sheep Mountains	Huerfano	37°41′N	105°11′W	> 9.7				Little Sheep Mountain 7½′	
Silver Mountain	Huerfano	37°36′N	105°06′W			> 0.6		La Veta 7½′	
Spanish Peaks	Huerfano and Las Animas	37°23′N	104°57′W			> 0.6		Spanish Peaks 7½′	
Sugar Loaf or Badito Peak	Huerfano	37°46′N	105°01′W					Badito Cone 7½′	

COUNTRY State, Group, and Laccolith Name	County or District	Latitude	Longitude	Length km	Width km	Thick- ness km	Depth of Intrusion km or fm	Topographic Map	References and Remarks
Tenmile group									
Chalk Mountain	Lake, Summit	39°23′N	106°12′W					Copper Mtn. 7 ½′	Emmons (1898)
Chicago Mountain	Eagle, Summit	39°24′N	106°15′W			0.6		Pando 7 ½′	Bergendahl and
Elk and Sugar Loaf Mountain	Eagle and Summit	39°28′N	106°15′W					Pando 7 ½′ Copper Mtn. 7 ½′	Koschmann (1971)
Gold Hill	Summit	39°25′N	106°09′W					Copper Mtn. 7 ½′	
Jacque Mountain	summit	39°27′N	106°11′W					Copper Mtn. 7 ½′	
McKinzie Gulch	Summit	39°29′N	106°09′W					Copper Mtn. 7 ½′	
West Elk group									
Anthracite Creek	Gunnison	38°55′N	107°12′W					Marcellina Mtn 7 ½′	Cross (1894)
Anthracite Range	Gunnison	38°49′N	107°08′W					Anthracite Range 7 ½′	Cady and Dapples (1937)
Chair Mountain or Mt. Sopris ?	Gunnison	39°01′N	107°15′W					Chair Mtn 7 ½′	Dapples (1940) Godwin and Gaskill (1964)
Crested Butte	Gunnison	38°53′N	106°56′W	4.4	3.22			Crested Butte 7 ½′	Gaskill and Godwin (1966)
Flat Top ?	Gunnison	38°42′N	106°54′W					Flat Top 7 ½′	Godwin (1968)
Gothic Mountain	Gunnison	38°57′N	107°00′W					Gothic and Oh-Be-Joyful 7 ½′	Gaskill and others (1981) Mutschler and others (1981)
Mt. Axtell	Gunnison	38°50′N	107°03′W					Mt. Axtell 7 ½′	Bevier and Klebanow (1984)
Mt. Beckwith	Gunnison	38°51′N	107°13′W					Anthracite Range 7 ½′	
Mt. Carbon	Gunnison	38°48′N	107°02′W					Mt. Axtell 7 ½′	
Mt. Gunnison	Gunnison	38°49′N	107°23′W					Minnesota Pass 7 ½′	
Mt. Marcellina	Gunnison	38°56′N	107°14′W	4.8		> 1.4		Marcellina Mtn 7 ½′	
Mt. Whetstone	Gunnison	38°49′N	106°58′W					Crested Butte 7 ½′	
Ragged Mountain	Gunnison	39°00′N	107°15′W					Chair Mtn 7 ½′	
Storm Ridge (Castle Pass)	Gunnison	38°47′N	107°11′W					Anthracite Range 7 ½′	
Treasure Mountain	Gunnison	39°01′N	107°02′W					Snowmass Mtn 7 ½′	
Connecticut									
Preston gabbro	New London	41°34′N	71°51′W					Jewett City 7 ½′	Loughlin (1912) Sclar (1958) Dixon (1982)
Hawaii									
Uwekahuna	Hawaii	19°24′N	155°16′W					Kipuka Pakekake 7 ½′	Powers (1916)

Additional references: Murata and Richter (1961), Aramiki (1968), MacDonald (1972)

Idaho									
Snake River Plain group									
Big Southern Butte	Butte	43°25′N	113°02′W					Big Southern Butte 7 ½′	MacDonald (1972) Spear (1977)
Blackfoot	Bingham	43°7′N	112°18′W					Buckskin Basin 7 ½′	Leeman and Gettings (1977)

Additional references: Karlo and Jorgenson (1979), Spear and King (1982)

Kansas									
Neosho Falls dome	Woodson	37°56′N	95°33′W					Piqua 7 ½′	Twenhofel (1926)
Rose dome	Woodson	37°48′N	95°42′W			0.43		Rose 7 ½′	Twenhofel and Bremer (1928)
Silver City dome	Woodson	37°45′N	95°47′W					Toronto SE and Middletown 7 ½′	Knight and Landes (1932)

Maine									
Mt. Desert group									
Bar Island	Hancock	44°24′N	68°12′W					Bar Harbor 15′	Chadwick (1942, 1944)
Bald Porcupine	Hancock	44°23′N	68°11′W					Bar Harbor 15′	Chapman (1969)
Burnt Porupine	Hancock	44°24′N	68°10′W					Bar Harbor 15′	
Cadillac Mountain	Hancock	44°15′N	68°15′W					Bar Harbor 15′	
Great Head	Hancock	44°20′N	68°10′W					Bar Harbor 15′	
Ireson Hill	Hancock	44°27′N	68°16′W					Mt. Desert 15′	
Long Porcupine	Hancock	44°24′N	68°10′W					Bar Harbor 15′	Feeder dike exposed.
Sheep Porcupine	Hancock	44°24′N	68°11′W					Bar Harbor 15′	

COUNTRY State, Group, and Laccolith Name	County or District	Latitude	Longitude	Length km	Width km	Thick- ness km	Depth of Intrusion km or fm	Topographic Map	References and Remarks
Sol's Cliff	Hancock	44°22'N	68°11'W					Bar Harbor 15'	

Minnesota-Wisconsin

Lake Superior group

Bad River	Ashland, Iron	46°30'N	90°30'W	96?	8?			Little Girls Point, Odanah, and Mellen 15' Saxon and Iron Belt 7½'	Lawson (1893) Bayley (1894, 1895) Van Hise and Leith (1911), Daly (1914) Grout (1918)
Beaver Bay	Lake	47°15'N	91°16'W						Schwartz and Davidson (1952)
Duluth	St. Louis, Lake	47°30'N	91°15'W	> 80	42	2.8			

Mississippi

Dubard and Whitaker	Grenada	33°46'N	89°58'W					Grenada 15'	Harrelson (1981)

Montana

Bearpaw Mountains group

Barber Butte ?	Blaine	48°17'N	109°25'W	2.4	1.2			Loyd 15'	Weed and Pirsson (1896a)
W. of Barber Butte?	Blaine	48°17'N	109°26'W	1.92	1.92			Loyd 15'	Reeves (1924, 1925)
Black Butte	Blaine	48°13'N	109°12'W	1.65	1.0			Rattlesnake 15'	Pecora (1941)
E. of Black Butte	Blaine	48°13'N	109°10'W	2.5	1.3			Rattlesnake 15'	Pecora and others
Black Joe Mtn. ?	Hill	48°09'N	109°44'W	3.22	3.22			Warrick 15'	(1957a, 1957b)
Boxelder	Hill	48°20'N	109°55'W	3.22	2.40			Laredo 15'	Kerr and others (1957)
E of Cleveland	Blaine	48°17'N	109°08'W			> 0.3		Cleveland 15'	Balsley and others
Cow and Gap Creeks	Blaine	48°06'N	109°11'W	1.3	1.0			Rattlesnake 15'	(1957a, 1957b, 1957c,
Dog Butte?	Blaine	48°16'N	109°05'W	1.5	1.5			Cleveland 15'	1957d), Stewart and
W. of Dog Butte?	Blaine	48°16'N	109°06'W	1.5	1.5			Cleveland 15'	others (1957), Bryant
Hanson Butte dome?	Blaine	48°11'N	109°09'W	1.8	1.6			Rattlesnake 15'	and others (1960),
Hanson Creek dome	Blaine	48°09'N	109°09'W	1.6	1.4	0.6		Rattlesnake 15'	Schmidt and others (1961)
Johnson Butte	Blaine	48°12'N	109°12'W	1.65	1.65			Rattlesnake 15'	Books (1962)
McCann Butte?	Blaine	48°17'N	109°04'W					Cleveland 15'	Schmidt and others (1964)
N. of McDonald Basin	Blaine	48°13'N	109°07'W	1.1	1.1			Rattlesnake 15'	Hearn and others (1964)
Miles Butte	Blaine	48°16'N	109°03'W	1.40	1.25			Cleveland 15'	Wager and Brown (1967)
Myrtle Butte	Blaine	48°19'N	109°03'W					Rattlesnake 15'	Peterson and Rambo
NW of Myrtle Butte	Blaine	48°15'N	109°14'W	1.1				Rattlesnake 15'	(1967, 1972)
W of Myrtle Butte	Blaine	48°14'N	109°06'W	1.8	1.65			Rattlesnake 15'	Kuhn (1982, 1983)
Peoples Creek	Blaine	48°13'N	109°13'W	1.8	1.8			Rattlesnake 15'	Leppert (1985)
Ryan's Butte	Chouteau	48°05'N	109°52'W	2.4	1.6			Centennial 15'	
Scotty Butte	Blaine	48°11'N	109°14'W	1.65	1.65			Rattlesnake 15'	
NE of Scotty Creek	Blaine	48°12'N	109°13'W	0.9	0.9			Rattlesnake 15'	
Snake Butte	Blaine	48°23'N	108°50'W	4.5	1.8	0.1	Bearpaw sh	Fort Belknap Agency 7½'	
Sucton Butte?	Blaine	48°12'N	109°06'W	3.1	1.7			Rattlesnake 15'	
Taylor Butte	Blaine	48°13'N	109°14'W	2.5	2.2			Rattlesnake 15'	
Timber Butte	Blaine	48°15'N	109°15'W	1.9	1.25	> 0.4		Cleveland 15'	
NNE of Timber Butte	Blaine	48°16'N	109°14'W	2.4	1.6	> 0.5		Cleveland 15'	

Beartooth Mountains group

Henderson Mountain	Park	45°03'N	109°57'W					Cooke City 15'	Rouse and others (1937)
Iron Mountain or Gold Hill	Sweetgrass	45°23'N	110°04'W					Mt. Douglas 15'	Parsons (1942) Parsons and Stow (1942)
Limestone Butte	Stillwater	45°28'N	109°53'W					Mt. Wood 15'	Foose and others (1961)
Line Creek Plateau	Carbon	45°03'N	109°22'W					Mt. Wood 15'	Hess (1960)
Meyers Canyon	Sweetgrass	45°28'N	109°58'W					Mt. Wood 15'	Horrall (1967)
North Meyers Canyon	Sweetgrass	45°30'N	110°00'W					Mt. Douglas 15'	Page (1977)
Round Mountain	Sweetgrass and Stillwater	45°27'N	109°56'W					Mt. Wood 15'	
NE of Round Mt.	Stillwater	45°28'N	109°55'W					Mt. Wood 15'	
Squaw Peak or Lodgepole	Sweetgrass	45°31'N	110°01'W					Mt. Douglas 15'	
Stillwater lopolith	Sweetgrass	45°20'N	110°10'W	48		5.5		Mt. Douglas, Mt. Wood, and Mt. Cowen 15' Mt. Rae, McLeod Basin, and Emerald Lake 7½'	

COUNTRY State, Group, and Laccolith Name	County or District	Latitude	Longitude	Length km	Width km	Thickness km	Depth of Intrusion km or fm	Topographic Map	References and Remarks
Big Belt Mountains group									
Birdtail Butte	Cascade	47°20′N	111°57′W	0.8	0.3	> 0.12		Simms 15′	Lyons (1944)
Cascade Butte	Cascade	47°18′N	111°44′W	5	3.5			Cascade 15′	Beall (1972, 1973)
Crown Butte	Cascade	47°26′N	111°56′W	2	1.5	> 0.18		Simms 15′	Beall (1977)
Fishback Butte	Cascade	47°20′N	111°56′W	0.8		> 0.12		Simms 15′	Whiting (1977a, 1977b)
Haystack Butte	Cascade	47°20′N	111°56′W	0.4		> 0.12		Simms 15′	McBride (1979)
Lionhead Butte	Cascade	47°20′N	111°59′W					Simms 15′	Hyndman and Alt (1982, 1987)
St. Peter	Cascade	47°18′N	111°56′W					Simms 15′	
Shaw Butte	Cascade	47°28′N	111°51′W	6	4	> 0.18	Eagle ss	Simms 15′	
Square Butte	Cascade	47°26′N	111°45′W	3	2.5			Simms and Cascade 15′	

Feeder dikes are reported to be visible for all laccoliths in this group.

Big Snowy Mountains group									
Big Snow Mtn. ?	Fergus and Golden Valley	46°47′N	109°27′W					Jump Off Peak 7½′	Reeves (1931)
Button Butte dome	Fergus	46°58′N	108°45′W						

Crazy Mountains group									
Coffin Butte	Meagher	46°22′N	110°07′W					Lebo Lake 7½′	Iddings and Weed (1894)
Comb Creek	Meagher	46°20′N	110°31′W					Rimrock Divide 7½′	Wolff (1938)
Gordon Butte	Meagher	46°25′N	110°21′W	4.3	4.2	0.27	Fort Union	Martinsdale 7½′	Bonini and others (1967)
Great Theralite Cliff?	Meagher	46°13′N	110°29′W	0.7		0.06		Virginia Peak 7½′	Larsen and Simms (1972)
Meagher	Meagher	46°25′N	110°22′W					Martinsdale 7½′	Emmart (1985)
Porcupine Butte?	Sweetgrass	46°10′N	110°06′W					Porcupine Butte 7½′	
Target Peak?	Meagher	46°11′N	110°23′W			> 0.03		Virginia Peak 7½′	
Three Peaks (Goat) Mountain	Meagher and Gallatin	46°10′N	110°34′W					Scab Rock Mountain 7½′	
Virginia Peak	Meagher	46°13′N	110°27′W	5		0.03		Virginia Peak 7½′	

Elkhorn Mountains group									
Black Butte stock?	Jefferson	46°17′N	111°57′W					Clancy 15′	Weed (1901)
Boulder batholith	Jefferson	46°15′N	112°15′W	110 +	60 +				Lawson (1914)
Buttleman	Broadwater	45°46′N	111°36′W	5.0		> 1.2	Flathead ss	Three Forks 15′	Klepper and others (1957)
Cemetary Ridge stock ?	Jefferson	46°16′N	111°56′W					Clancy 15′	Robinson (1963)
Elkhorn Peak?	Jefferson	46°18′N	111°55′W					Clancy 15′	Klepper and others (1971)
Jenkins Gulch dome	Jefferson	46°17′N	111°47′W					Clancy 15′	Wager and Brown (1967)
Milligan Cr. stock?	Jefferson	45°55′N	111°40′W					Three Forks 15′	
Queen Gulch	Jefferson	46°15′N	111°57′W					Clancy 15′	
Sagebrush Park stock?	Jefferson	46°11′N	111°52′W					Devils Fence 15′	
Spring Creek?	Jefferson	46°20′N	111°46′W					Clancy 15′	
Star Ranch dome	Jefferson	46°08′N	111°49′W	22	10	> 1.0		Devils Fence 15′	
10 N	Broadwater	45°59′N	111°34′W	> 6.4	4.8			Three Forks 15′	
Turnley Ridge stock?	Jefferson	46°15′N	111°57′W					Clancy 15′	

Highwood Mountains group									
Antelope	Chouteau	47°37′N	110°20′W					Geraldine 7½′	Weed and Pirsson
Cowboy Creek	Chouteau	47°30′N	110°18′W			> 0.05		Geraldine 7½′	(1895a, 1896a, 1901)
Geraldine	Chouteau	47°36′N	110°19′W				Eagle ss	Geraldine 7½′	Weed (1899a)
Lost Lake	Chouteau	47°39′N	110°28′W					Montague 7½′	Pirsson (1905)
Major Shonkin Sag	Chouteau	47°34′N	110°19′W	5	5	0.45	Eagle ss	Geraldine 7½′	Reeves (1929)
Minor Shonkin Sag	Chouteau	47°31′N	110°15′W				Eagle ss	Geraldine 7½′	Osborne and Roberts (1931)
Montague	Chouteau	47°38′N	110°27′W	2.4				Montague 7½′	Larsen and others (1935)
Palisade or Round Butte	Chouteau	47°28′N	110°19′W					Jiggs Flat 7½′	Hurlburt (1936)
Square Butte	Chouteau	47°28′N	110°14′W	5		0.4	Eagle ss	Square Butte 7½′	Barksdale (1937)

Additional references: Reynolds (1937), Hurlburt and Griggs (1939), Stephenson (1940), Larsen (1941), Barksdale (1950, 1952), Nash and Wilkinson (1969, 1970, 1971), Nash (1972), Nicoll and Nicholls (1974), Bhattacharya and Navolio (1975), Edmond (1980), Kendrick (1980a, 1980b), Kleinkopf (1980), Kendrick and Edmond (1981), Dockstader (1982), Nash and Kendrick (1982), Liptak (1984), Hirsch and Hyndman (1985)

COUNTRY State, Group, and Laccolith Name	County or District	Latitude	Longitude	Length km	Width km	Thickness km	Depth of Intrusion km or fm	Topographic Map	References and Remarks
Judith Mountains group									
Alpine	Fergus	47°08′N	109°15′W	5	3.2			Lewistown and Judith Peak 15′	Weed and Pirsson (1898)
Bald Butte	Fergus	47°13′N	109°16′W					Lewistown 15′	
Black Butte	Fergus	47°14′N	108°59′W						
Burnett Creek	Fergus	47°07′N	109°18′W					Judith Peak 15′	
Cone Butte	Fergus	47°15′N	109°05′W					Judith Peak 15′	
Crystal Peak	Fergus	47°10′N	109°13′W					Judith Peak 15′	
Deer Creek dome	Fergus	47°15′N	109°16′W	2				Lewistown 15′	
Elk Peak	Fergus	47°14′N	109°13′W					Judith Peak 15′	
Flat Mountain	Fergus	47°05′N	109°15′W					Lewistown 15′ and Judith Peak 15′	
Gold Hill stock?	Fergus	47°10′N	109°14′W					Judith Peak 15′	
Judith Peak	Fergus	47°14′N	109°14′W	8	>6.5			Judith Peak 15′	
Kelly Hill	Fergus	47°05′N	109°16′W					Lewistown 15′	
Lookout Peak	Fergus	47°15′N	109°02′W						
Maginnis anticline	Fergus	47°09′N	109°09′W	1.6	1.6			Judith Peak 15′	
Maginnis Peak?	Fergus	47°13′N	109°11′W					Judith Peak 15′	
Pyramid Gulch	Fergus	47°06′N	109°16′W					Lewistown 15′	
Pyramid Peak	Fergus	47°07′N	109°16′W					Lewistown 15′	
Ruby Gulch	Fergus	47°07′N	109°16′W					Lewistown 15′	
Warm Spring or Creek dome	Fergus	47°12′N	109°17′W					Lewistown 15′	
West Armell Creek	Fergus	47°14′N	109°15′W					Lewistown 15′	
Little Belt Mountains group									
Barker	Judith Basin	47°05′N	110°41′W	5.0		1.2		Barker 7½′	Weed and Pirsson (1896c)
Butcherknife Mt.	Judith Basin	47°01′N	110°34′W	7.2		>0.6		Mixes Baldy 7½′	Weed (1899b, 1900)
Castle Mountain	Meagher	46°30′N	110°45′W	12.8	7.2			Castle Town 7½′	Goodspeed (1946)
Clendinnin-Peterson	Judith Basin	47°06′N	110°36′W	8.0	3.2	1.1		Mixes Baldy 7½′	Witkind (1965b, 1966,
Dry Wolf	Judith Basin	47°00′N	110°30′W					Mixes Baldy 7½′ and Bandbox 7½′	1971), Kleinkopf and others (1972)
Granite Mountain	Judith Basin	47°06′N	110°30′W					Mixes Baldy 7½′	Witkind (1973)
Hoover Ridge dome	Judith Basin	47°00′N	110°42′W					Neihart, Barker 7½′	Corry (1976)
Limestone Butte	Judith Basin	47°10′N	110°41′W	6.4		>0.3	>0.6	Limestone Butte 7½′	
Mixes Baldy - Anderson Peak	Judith Basin	47°04′N	110°36′W	4.0		>0.6		Mixes Baldy 7½′	
Otter	Judith Basin	47°08′N	110°43′W	3.2		>0.2	Heath sh	Limestone Butte 7½′	
Skull Butte?	Judith Basin	47°05′N	110°15′W					Cayuse Basin 7½′	
Taylor Mountain	Judith Basin	47°04′N	110°32′W	6.4	4.0	0.54	Jefferson dolomite	Mixes Baldy 7½′	
Thunder Mountain	Judith Basin	47°02′N	110°53′W					Thunder Mtn. 7½′	
Little Rocky Mountains group									
Bull Creek dome	Phillips	47°52′N	108°45′W					Shetland Divide 7½′	Weed and Pirsson (1896b)
Cone Butte	Phillips	47°55′N	108°26′W					Bear Mountain 7½′	Knechtel (1959)
Crown Butte	Phillips	47°53′N	108°36′W					Hays SE 7½′	Bonini and others (1967)
Cyprian or Sipary Anne Butte?	Phillips	47°52′N	108°41′W					D-Y Junction 7½′	Roemmel (1982)
Eagle Child dome	Blaine	47°56′N	108°41′W					Hays 7½′	More than fifty domes exist to the
East Coburn Butte?	Phillips	47°53′N	108°22′W					Coburn Butte 7½′	south and some of them
Grouse-Alder dome	Phillips	47°51′N	108°31′W					Hays SE 7½′	have exposed igneous
Hartmann dome?	Phillips	47°55′N	108°26′W					Robinson School 7½′	cores.
Indian Head Peak or Indian Butte	Phillips	47°54′N	108°38′W					Hays 7½′	
Little Rocky Mtns.	Phillips	47°56′N	108°34′W					Zortman 7½′	
Lookout Butte	Phillips	47°53′N	108°35′W					Zortman 7½′	
McNeal dome	Blaine	47°57′N	108°40′W					Hays 7½′	
Morrison Butte	Phillips	47°51′N	108°35′W					Hays SE 7½′	
Monument dome	Blaine	48°01′N	108°36′W					Lodge Pole 7½′	
Powderface dome	Blaine	48°01′N	108°37′W					Stiffarm Coulee and Lodge Pole 7½′	
Saddle Butte or Castle Butte	Phillips	47°53′N	108°32′W					Zortman 7½′	
Spring Park dome	Phillips	47°55′N	108°30′W					Zortman 7½′	
Thorsen dome	Phillips	47°54′N	108°46′W					Crazyman Coulee 7½′	
West Coburn Butte?	Phillips	47°53′N	108°23′W					Bear Mountain 7½′	
White Cow	Blaine	48°01′N	108°38′W					Lodge Pole 7½′	

COUNTRY State, Group, and Laccolith Name	County or District	Latitude	Longitude	Length km	Width km	Thickness km	Depth of Intrusion km or fm	Topographic Map	References and Remarks
Moccasin Mountains group									
Block F ?	Fergus	47°09′N	109°31′W					Spring Creek Junction 7 ½′	Palmer (1925)
East Peak	Fergus	47°10′N	109°30′W	2.5	1.2		Heath	Spring Creek Junction 7 ½′	Blixt (1933) Miller (1959)
Hanover dome	Fergus	47°08′N	109°33′W	2.4	2.4	0.52	> Madison ls	Spring Creek Junction 7 ½′	Lindsey (1982) Lindsey and Naeser (1985)
Little Meadow Creek	Fergus	47°15′N	109°29′W					Spring Creek Junction 7 ½′	
North Moccasin Mt.	Fergus	47°16′N	109°29′W					Spring Creek Junction 7 ½′	
South Peak dome	Fergus	47°10′N	109°33′W	4.8	3.2	1.1	> Madison ls	Spring Creek Junction 7 ½′	
Tower Peak	Fergus	47°11′N	109°32′W	6.4		1.74		Spring Creek Junction 7 ½′	
Sweet Grass Hills group									
East Butte	Liberty	48°52′N	111°07′W					Hawley Hill, Haystack Butte, and Bingham Lake 7 ½′	Weed and Pirsson (1895b)
Gold Butte	Toole	48°52′N	111°19′W					Cameron Lake 7 ½′	
West Butte	Toole	48°56′N	111°33′W					West Butte 15′	
Nevada									
Goldfield	Esmeralda and Nye	37°44′N	117°11′W	10				Mud Lake 15′	Ransome (1909)
Humboldt	Churchill and Pershing	40°N	118°W	50	~25				Speed (1976)
New Hampshire-Maine									
Exeter ?		43°15′N	70°55′W	32	7	3?		Berwick and Dover 15′	Bothner (1974)

Gravity study indicates +15 mGal anomaly over diorites.

Oliverian group									
Croydon dome	Sullivan	43°27′N	72°13′W	19.2				Ammonoosuc volcanics	Mascoma and Sunapee 15′	Chapman (1939)
Lebanon dome?	Grafton	43°40′N	72°15′W				Post Pond volcanics	Hanover and Mascoma 15′	Chapman (1942) Kane and Brommery (1968), Naylor (1968)
Mascoma dome	Grafton	43°10′N	72°06′W					Mt. Cube and Mascoma 15′	Leo (1979)
Mt. Clough ?	Grafton	43°45′N	72°W	> 104				Mt. Cardigan 15′	Leo (1980a, 1980b)
Smarts Mountain dome	Grafton	43°45′N	72°03′W					Mt. Cube and Mascoma 15′	Foland and Loiselle (1981)
Unity dome	Sullivan	43°20′N	72°17′W	16			Ammonoosuc volcanics	Claremont 15′	

This is an incomplete list of the Oliverian domes, and there is debate as to whether any, or all of these domes are laccoliths.

New Mexico									
Abo Pass	Torrance	34°25′N	106°20′W					Abo 7 ½′	Sanford and others (1977) Brocher (1978)
Cuchillo Mountain	Sierra	33°28′N	107°36′W					Jaralosa Mtn 7 ½′	McMillan (1979) McMillan and Jahns (1979)
Wind Mountain	Otero	32°03′N	105°30′W					Cornudas Mtn and McVeigh 7 ½′	Burton and Jackson (1979)
Gallinas Mountains group									
Cougar Mountain?	Torrance and Lincoln	34°15′N	106°43′W					Becker SW 7 ½′	Kelley (1946) Elston and Snider (1964)
Gallinas Peak	Torrance and Lincoln	34°15′N	106°47′W					La Joya 7 ½′	Perhac (1964)
Red Cloud	Lincoln	34°13′N	106°45′W					Mesa del Yeso 7 ½′	

COUNTRY State, Group, and Laccolith Name	County or District	Latitude	Longitude	Length km	Width km	Thickness km	Depth of Intrusion km or fm	Topographic Map	References and Remarks
Sierra del Oro group									
Los Cerrillos	Santa Fe	35°28′N	106°08′W					Madrid 15′	Keyes (1918)
Ortiz	Santa Fe	35°20′N	106°10′W					Madrid 15′	Keyes (1922)
San Pedro or Tuertos	Santa Fe	35°15′N	106°12′W					Edgewood, Madrid 15′	Thompson (1964)
South Mountain or San Ysidro	Santa Fe	35°12′N	106°12′W					San Pedro 7½′	
Silver City group									
Allie Canyon	Grant	32°55′N	108°02′W					Allie Canyon 7½′	Paige (1916)
Big Burro Mountains	Grant	32°36′N	108°24′W	8				Burro Peak 7½	Jones and others (1961)
Copper Flat dome	Grant	32°48′N	108°07′W					Santa Rita 7½′	Pratt and Jones
Fort Bayard	Grant	32°47′N	108°09′W					Fort Bayard 7½′	(1961, 1965)
Gomez Peak	Grant	32°51′N	108°17′W					Silver City 7½′	Jones and others (1967)
Hanover-Fierro anticline	Grant	32°50′N	108°05′W					Santa Rita 7½′	Aldrich (1974) Peterson (1979)
Kneeling Nun	Grant	32°47′N	108°02′W					Santa Rita 7½′	Kolessar (1970, 1982)
Lone Mountain or Cameron Creek	Grant	32°43′N	108°10′W					Hurley West 7½′	Simmons (1984)
Pinos Altos	Grant	32°52′N	108°13′W					Fort Baynard 7½′	
Santa Rita	Grant	32°47′N	108°06′W					Santa Rita 7½′	
Silver City	Grant	32°46′N	108°16′W	2.4				Silver City 7½′	
Whisky Creek?	Grant	32°50′N	108°13′W					Fort Bayard 7½′	
Whitewater Creek	Grant	32°43′N	108°08′W					Hurley West 7½′	
Turkey Mountains group									
								Maxson Crater, Optimo, La Chata Crater, and Cerro Negro 7½′	Laccoliths are present in the Turkey Mt.-ains, but at present number and precise locations are unknown.
New York									
Adirondacks group									
Butterfield Lake	St. Lawrence and Jefferson	44°21′N	75°50′W					Alexandria Bay and Hammond 15′	Alling (1919) Miller (1929)
California	St. Lawrence	44°14′N?	75°25′W?					Lake Bonaparte 7½′	Buddington (1929)
Canton	St. Lawrence	44°34′N?	75°09′W?					Canton 7½′	Cannon (1937)
Clarks Pond	St. Lawrence	44°16′N?	75°16′W?					Edwards 7½′	Buddington (1939)
Dodds Creek	St. Lawrence	44°16′N?	75°38′W?					Muskellunge Lake 7½′	Buddington (1948)
Diana	Lewis	44°00′N	75°20′W					Remington Corners 7½′	de Waard (1961) Engel and Engel (1963)
Edwards	St. Lawrence	44°20′N	75°15′W					South Edwards and Edwards 7½′	
Fish Creek	St. Lawrence	44°30′N	75°30′W					Hammond 15′ and Richville, Gouverneur 7½′	
Gouverneur	St. Lawrence	44°22′N	75°24′W					Gouverneur 7½′	
Hermon	St. Lawrence	44°25′N?	75°12′W?					Hermon 7½′	
Hickory Lake	St. Lawrence	44°24′N	75°30′W					Pope Mills 7½′	
Hyde	Jefferson	44°10′N	75°49′W					Theresa 7½′	
Jay-Mt. Whiteface	Essex	44°20′N	73°40′W					Au Sable Forks 15′	
Payne Lake	Jefferson	44°17′N	75°38′W					Muskellunge Lake 7½′	
Piseco Lake	Hamilton	43°25′N	74°35′W					Piseco Lake 15′	
Pyrites	St. Lawrence	44°30′N	75°13′W					Canton 7½′	
Reservoir Hill	St. Lawrence	44°20′N	75°20′W					Gouverneur 7½′	
Rossie	St. Lawrence	44°23′N	75°38′W					Hammond 7½′	
Santa Clara	Franklin	44°38′N	74°28′W					Meno and Santa Clara 7½′	
Thirteenth Lake	Warren	43°43′N	74°08′W					Thirteenth Lake 15′	
Tupper-Saranac	Franklin	44°15′N	74°25′W					St. Regis 15′ and Long Lake 15′	
NW of Westport	Essex	44°16′N	73°29′W					Willsboro 15′	
Oklahoma									
Wichata lopolith	Greer	34°55′N	99°25′W					Granite 7½	Hamilton (1959)
Oregon									
Bald Mountain?	Josephine	42°40′N	123°21′W					Glendale 15′	Diller (1903)

COUNTRY State, Group, and Laccolith Name	County or District	Latitude	Longitude	Length km	Width km	Thickness km	Depth of Intrusion km or fm	Topographic Map	References and Remarks
									Daly (1914) Gabbro intrusives
Coxine Creek?	Yamhill	45°12′N	123°16′W					Sheridan 15′	Baldwin and others (1955)
Stony Mountain	Yamhill	45°14′N	123°26′W			0.15		Sheridan 15′	

Gabbro intrusives of apparent laccolithic character. Other laccoliths may be present in this area.

COUNTRY State, Group, and Laccolith Name	County or District	Latitude	Longitude	Length km	Width km	Thickness km	Depth of Intrusion km or fm	Topographic Map	References and Remarks
Hinckel Creek	Douglas	43°26′N	123°04′W	5		0.3		Glide 15′	Daly (1914) Gabbro intrusives
Roseburg?	Douglas	43°12′N	123°20′W	>40				Roseburg 15′	Diller (1898)
Willow Lake?	Baker	44°51′N	118°05′W						Taubeneck and Poldervaart (1960)

South Dakota-Wyoming

Black Hills group

COUNTRY State, Group, and Laccolith Name	County or District	Latitude	Longitude	Length km	Width km	Thickness km	Depth of Intrusion km or fm	Topographic Map	References and Remarks
Annie Creek or Spotted Tail dome	Crook	44°21′N	104°15′W					Tinton, Sundance 15′	Irving (1899)
Bald Mountain	Crook	44°16′N	104°05′W					Tinton 15′	Jaggar (1901, 1904) Darton (1905, 1909)
Bear Butte	Meade	44°28′N	103°25′W			>0.6	Minnelusa ss	Ft. Meade 7½′	Darton and Paige (1925)
Bear Den Mountain	Lawrence	44°20′N	103°37′W			>0.1	Minnekahta ls	Deadman Mtn 7½′	Runner (1943)
Black Buttes	Crook	44°17′N	104°16′W				Deadwood and Spearfish	Sundance 15′	Honkala (1949) Brown (1952)
Black Hills dome?		44°N	104°W	400	130				Brown and Lugn (1952)
Burro Gulch	Lawrence	44°24′N	103°49′W					Spearfish 7½′	Brown (1956)
Cement Ridge	Crook	44°23′N	104°05′W					Tinton 15′	Waagé (1959)
Circus Flats	Meade	44°28′N	103°27′W					Ft. Meade 7½′	Mapel and others (1959)
Citadel Rock	Lawrence	44°26′N	103°56′W					Maurice 7½′	Sotteck (1959)
Crook Mountain	Lawrence	44°25′N	103°37′W					Deadwood West and Sturgis 7½′	Basset (1961) Robinson and others (1964)
Crow Peak	Lawrence	44°28′N	103°57′W	4.0				Maurice 7½′	Getz (1965)
Custer Peak	Lawrence	44°15′N	103°44′W	>3			Deadwood	Minnesota Ridge 7½′	Mukhurjee (1968)
Cutting stock?	Lawrence	44°22′N	103°49′W						Fisher (1969)
Deadman Mountain	Meade	44°23′N	103°32′W			>0.3		Deadman Mtn 7½′	Kleinkopf (1970)
Deadman Mountain or Boulder Park	Lawrence	44°24′N	103°35′W					Sturgis 7½′	Grunwald (1970) Kirchner (1971)
Deer Mountain	Lawrence	44°18′N	103°48′W				Deadwood	Lead 7½′	Anna (1973)
Dome Mountain	Lawrence	44°20′N	103°40′W			0.15	Deadwood	Deadwood South 7½′	Rockey (1974)
Elkhorn Peak	Lawrence	44°28′N	103°42′W					Deadwood North 7½′	Kleinkopf and Redden (1975)
Elk Mountain	Lawrence	44°21′N	103°53′W			>0.05		Savoy 7½′	Beck (1976)
Green Mountain	Crook	44°24′N	104°19′W	3.5	3.5	>0.2	Minnelusa ss	Sundance 15′	Elwood (1976)
Green Mountain or Polo Peak	Lawrence	44°25′N	103°44′W					Deadwood North 7½′	Heidt (1977) Jumnogthai (1979)
Homestake	Lawrence	44°22′N	103°47′W					Lead 7½′	Matthews (1979)
Inyan Kara Mountain	Crook	44°13′N	104°21′W					Inyan Kara Mountain 7½′	Meier (1979) Sofranoff (1979)
Lime Butte	Crook	44°21′N	104°22′W	~2.0			Minnekahta	Sundance 15′	Usiriprisan (1979)
Little Missouri Buttes	Crook	44°37′N	104°48′W					Oshoto 15′	Halvorson (1980)
Lytle Creek	Crook	44°32′N	103°29′W					Alva 15′	White (1980)
Mato Tepee or Devil's Tower	Crook	44°36′N	104°44′W	~3.0		>0.3		Devils Tower 15′	Beck and Lisenbee (1981) Heidt (1981)
Mt. Theo. Roosevelt or Sheep Mountain	Custer	44°20′N	103°45′W					Deadwood South and Lead 7½′	Lisenbee (1981) Matthews (1981)
Needles	Lawrence	44°24′N?	104°01′W?					Tinton 15′	Sofranoff (1981)
Nigger Hill	Crook and Lawrence	44°23′N	104°02′W	>10			Deadwood	Sundance 15′	Rich (1985)
Pillar Peak	Lawrence	44°20′N	103°40′W			0.16		Deadwood South 7½′	
Ragged Top	Lawrence	44°22′N	103°53′W			>0.1		Savoy 7½′	
Richmond Hill	Lawrence	44°23′N	103°50′W					Spearfish 7½′	
Spearfish Peak	Lawrence	44°26′N	103°51′W			>0.1	Pahasapa ls	Spearfish 7½′	
Strawberry Mt.	Weston	44°10′N	104°16′W	~3.0	~1.8		>Pahasapa	Inyan Kara Mtn 7½′	
Strawberry Ridge or Spruce Gulch	Lawrence	44°22′N	103°42′W			0.12		Deadwood South 7½′	
Sugarloaf Mountain	Lawrence	44°19′N	103°47′W			~0.1	Deadwood	Savoy 7½′	
Sundance	Crook	44°23′N	104°22′W	1.5	0.8			Sundance 15′	
Terry Peak	Lawrence	44°20′N	103°50′W			>0.1	Deadwood	Lead 7½′	
Tetro Rock	Lawrence	44°25′N	103°48′W			>0.1		Spearfish 7½′	
Tilford	Lawrence	44°18′N	103°28′W					Tilford 7½′ and Minnelusa ss	
Twin Peaks	Lawrence	44°22′N	103°52′W					Spearfish and Lead 7½′	

COUNTRY State, Group, and Laccolith Name	County or District	Latitude	Longitude	Length km	Width km	Thick-ness km	Depth of Intrusion km or fm	Topographic Map	References and Remarks
Vanocker or Kirk Hill	Lawrence	44°19′N	103°34′W			>0.3	Deadwood and Minnelusa ss	Deadman Mtn 7½′	
War Eagle Hill	Lawrence	44°21′W	103°50′W					Lead 7½′	
Warren Peak or Bearlodge Mountains	Crook	44°28′N	104°27′W	32.0	>13.0			Sundance 15′	
Whitewood Peak	Lawrence	44°24′N	103°39′W				Deadwood	Deadwood North 7½′	
Woodville Hills	Lawrence	44°17′N	103°46′W	>2		>0.12	Deadwood	Lead 7½′	
Texas									
Diablo Plateau	Otero and Hudspeth	32°00′N	105°30′W					Cornudas Mtn 7½′ and San Antonio Mtn 15′	Barker and others (1977)
Hueco Mountains	El Paso and Hudspeth	31°50′N?	106′W?					Hueco Mtn 15′	Wise (1977)
Wolf Mountain	Llano	30°49′N	98°41′W	16	8.8			Llano North 7½′	Stenzel (1936)
Balcones group									
Mustang Hill or Pinto Mountain?	Uvalde	29°18′N	100°04′W	2.4		0.12	Eagle Ford	Mustang Waterhole 7½′	Greenwood and Lynch (1959)
Turkey Mountain?	Kinney	29°22′N	100°12′W					Salmon Peak 7½′	
	About 90 small intrusions are reported along the Balcones fault zone in central Texas. No information is available on structure of these intrusions. Mustang Hill is basalt with estimated density contrast of 500 kg per cubic meter.								
Big Bend group (also continues south into Coahuila,Mexico)									
Agua Fria	Brewster	29°31′N	103°39′W	3.6?				Agua Fria 7½′	Powers (1921)
Bee Mountain	Brewster	29°21′N	103°32′W	1.1?				Terlingua 7½′	Ross (1937)
Black Mesa	Brewster	29°21′N	103°44′W	2.41		0.18		Marilla Mtn 7½′	Lonsdale (1940)
Bogles domes	Presidio	29°28′N	103°53′W	2.15		0.61		Sauceda Ranch 7½′	Herrin (1957)
Chilcotal Mountain?	Brewster	29°12′N	103°09′W	8.1?	2.0?			Glenn Spring 7½′	Yates and Thompson (1959)
Chisos Pen anticline?	Brewster	29°19′N	103°24′W					Tule Mt. 7½′	Maxwell and others (1967)
Christmas Mountains	Brewster	29°26′N	103°26′W					Christmas Mt. 7½′	McKnight (1970)
Cigar Mountain	Brewster	29°19′N	103°35′W	1.6?				Terlingua 7½′	Corry (1972, 1976)
Contrabando	Presidio and Brewster	29°18′N	103°47′W	>3.2		0.15		Lajitas 7½′	Kovschak (1974, 1976) Lewis (1976)
Contrabando Lowlands	Presidio	29°18′N	103°52′W	7.7		0.31		Lajitas 7½′	Indest and Carman (1979)
Corrazones Peaks	Brewster	29°30′N	103°24′W					Christmas Mt. 7½′	Bagstad (1981)
Croton-Paint Gap Hills	Brewster	29°22′N	103°20′W	8.1	3.2			Sombrero Peak and The Basin 7½′	Shepard (1982a, 1982b) Mosconi (1984)
Cow Heaven anticline?	Brewster	29°06′N	103°15′W					Reed Camp 7½′	Henry and others (1986)
Cuesta Blanca	Brewster	29°18′N	103°34′W	>3.2				Terlingua 7½′	Corry and others (1988)
Dominguez Mountain	Brewster	29°09′N	103°18′W	2.8?	1.6?			Emory Peak 7½′	
Fossil Knobs	Brewster	29°20′N	103°37′W	2.01	0.80			Terlingua 7½′	
Glenn Springs	Brewster	29°11′N	103°10′W	1.6?	1.6?			Glenn Spring 7½	
Government Spring	Brewster	29°20′N	103°17′W	3.2?	1.6?			The Basin 7½′	
Grapevine Hills	Brewster	29°24′N	103°13′W	3.22	3.22			Grapevine Hills 7½′	
Gray Hill	Brewster	29°34′N	103°22′W	1.6?				Agua Fria 15′	
Indian Head	Brewster	29°21′N	103°30′W	~2				Terlingua 7½′	
Leon Mountain	Brewster	29°34′N	103°21′W	1.6?	0.8?			Terlingua 7½′	
Little dome	Presidio	29°24′N	103°55′W					Sauceda Ranch 7½′	
Llano dome	Presidio	29°27′N	103°55′W					Sauceda Ranch 7½′	
Long Draw	Brewster	29°22′N	103°40′W	4.83		>0.09		Marilla Mtn 7½′	
Mariscal Mountain	Brewster	29°01′N	103°09′W					Mariscal Mt. 7½′	
Maverick Mountain	Brewster	29°19′N	103°30′W	2.4?	1.2?			Terlingua 7½′	
McKinney Hill	Brewster	29°24′N	103°03′W	9.66	4.83			McKinney Springs 7½′	
Nine Point Mesa?	Brewster	29°38′N	103°28′W					Nine Point Mesa 7½′	
Packsaddle	Brewster	29°31′N	103°34′W	1.1?	0.8?			Agua Fria 15′	
Panther	Brewster	29°17′N	103°14′W	3.22		1.6		Panther Junction 7½′	
Panther dome	Presidio	29°22′N	103°57′W					Santana Mesa 7½′	
Panther Mountain	Brewster	29°31′N	103°36′W	1.6?				Agua Fria 15′	
Rattlesnake Mtn.	Brewster	29°16′N	103°33′W	2.4?				Castolon and Terlingua 7½′	
Red Bluff	Brewster	29°35′N	103°30′W	1.3?	0.8?			Agua Fria 15′	
Rosillos Mountain	Brewster	29°28′N	103°14′W	>10	>10	>1		Grapvine Hills, Sombrero Peak, and Bone Spring 7½′	
San Vicente anticline?	Brewster	29°10′N	103°04′W					San Vicente 7½′	
Saucita dome	Presidio	29°29′N	103°57′W	4.6	0.92	0.61		Sauceda Ranch 7½′	
Sawmill Mountain	Brewster	29°22′N	103°37′W	1.8?				Terlingua 7½′	
Segundo dome	Presidio	29°25′N	103°54′W					Sauceda Ranch 7½′	
Sierra Quemada?	Brewster	29°11′N	103°19′W	9.7?				Emory Peak 7½′	

COUNTRY State, Group, and Laccolith Name	County or District	Latitude	Longitude	Length km	Width km	Thickness km	Depth of Intrusion km or fm	Topographic Map	References and Remarks
Sierra San Vicente	Brewster and Coahuila, Mexico	29°05′N	103°04′W					Solis 7½′	Bulk of laccolith is in Mexico.
Small Pox Well	Brewster	29°26′N	103°21′W	1.1?		0.04?		Sombrero Peak 7½′	
Solitario	Presidio and Brewster	29°17′N	103°48′W	12.0	12.0	1.6	2.0	Solitario 7½′	
Study Butte	Brewster	29°18′N	103°32′W					Terlingua 7½′	
Talley Mountain	Brewster	29°08′N	103°10′W	1.6?				Glenn Spring 7½′	
Tapado dome	Presidio	29°24′N	104°04′W					Agua Adrento Mtn 7½′	
Terlingua anticline	Brewster	29°21′N	103°45′W	~16		1.0?		Yellow Hill, Solitario, and Marilla Mt. 7½′	
Tortuga Mountain	Brewster	29°11′N	103°16′W	3.2	1.6?			Emory Peak 7½′	
248	Brewster	29°20′N	103°33′W	3.2?		0.15		Terlingua 7½′	
Ward-Pulliam Mtn.	Brewster	29°16′N	103°20′W	8.1?	3.6?			The Basin 7½′	
Wax Factory	Presidio	29°20′N	103°49′W	3.22		0.05		Lajitas 7½′	
Wildhorse Mountain	Brewster and Hudspeth	29°23′N	103°31′W					Hen Egg Mtn 7½′	
Willow Mountain?	Brewster	29°22′N	103°31′W	~2	~1.6			Terlingua 7½′	

Quitman and Sierra Blanca Mountains group

Little Blanca Mountain	Hudspeth	31°17′N	105°26′W					Triple Hill 15′	Huffington (1943)
Little Round Top	Hudspeth	31°17′N	105°27′W					Triple Hill 15′	Albritton and Smith
Rainy and Pinnacle Peaks	Hudspeth	31°11′N	105°30′W					Sierra Blanca and Dome Peak 7½′	(1965), Wiley (1972) McAnulty (1980),
Sierra Blanca	Hudspeth	31°15′N	105°26′W					Sierra Blanca 7½′ and Triple Hill 15′	Shannon and Goodell (1983)
Round Top	Hudspeth	31°22′N	105°28′W					Triple Hill 15′	
Triple Hill	Hudspeth	31°17′N	105°23′W					Triple Hill 15′	

Utah

Abajo Mountains group

Abajo Peak	Grand	37°52′N	109°28′W	6.4	1.6	.13-.82	Mancos sh	Verdure 2 NW and Verdure 2 SW 7½′	Thorpe (1919, 1938) Witkind (1957, 1958,
Allen Canyon	Grand	37°49′N	109°34′W	2.56	0.8	0.18	Morrison	Oak Ridge 1 SE 7½′	1964b, 1964c, 1965a)
Arrowhead	Grand	37°47′N	109°28′W	1.6	1.44	0.30	Morrison	Verdure 2 SW 7½′	Case and Joesting (1972)
Camp Jackson	Grand	37°48′N	109°29′W	3.2	1.92	.24-.48	Morrison	Verdure 2 SW 7½′	Witkind (1975)
Corral anticline	Grand	37°53′N	109°29′W	1.3	1.0	0.26	Morrison?	Verdure 2 NW 7½′	
Dickson Creek	Grand	37°48′N	109°26′W	0.96	0.96	0.13	Dakota ss and Burro Canyon	Verdure 2 SW 7½′	
Dry Wash	Grand	37°49′N	109°32′W	0.8	0.8	0.18	Morrison	Oak Ridge 1 SE 7½′	
Gold Queen	Grand	37°51′N	109°26′W	3.2	3.04	0.45	Morrison	Verdure 2 SW 7½′	
Horsehead	Grand	37°52′N	109°28′W	2.74	2.08	.78	Dakota ss and Burro Canyon	Verdure 2 NW 7½′ and Verdure 2 SW 7½′	
Indian Creek	Grand	37°50′N	109°32′W	2.22	1.44	0.61	Morrison	Oak Ridge 1 SE 7½′	
Jackson Ridge	Grand	37°51′N	109°29′W	2.53	.96	.43	Mancos sh	Elk Ridge 1 SE 7½′ and Verdure 2 SW 7½′	
Knoll	Grand	37°48′N	109°27′W	0.8	0.48	0.10	Morrison	Verdure 2 SW 7½′	
Locust	Grand	37°49′N	109°28′W	.96	.32	.18	Dakota ss and Burro Canyon	Verdure 2 SW 7½′	
Mt. Linnaeus	Grand	37°49′N	109°33′W	1.44	1.28	0.37	Mancos sh	Oak Ridge 1 SE 7½′	
North Flank	Grand	37°57′N	109°32′W	1.12	0.32	0.18	Morrison	Oak Ridge 1 NE 7½′	
North Robertson Pasture	Grand	37°53′N	109°30′W	2.22	2.08	.54	Morrison	Elk Ridge 1 NE 7½′ and Verdure 2 NW 7½′	
Pole Creek	Grand	37°52′N	109°26′W	1.6	1.0	.3	Dakota ss and Burro Canyon	Verdure 2 NW 7½′	
Porcupine	Grand	37°48′N	109°31′W	2.08	1.12	0.78	Morrison	Oak Ridge 1 SE 7½′	
Prospect	Grand	37°48′N	109°32′W	2.08	1.12	0.43	Cutler	Oak Ridge 1 SE 7½′	
Ridge	Grand	37°49′N	109°30′W	0.64	0.64	0.24	Morrison	Oak Ridge 1 SE 7½′	
Rocky Trail	Grand	37°47′N	109°28′W	0.8	0.48	0.18	Morrison	Verdure 2 SW 7½′	
Scrub oak	Grand	37°48′N	109°31′W	1.28	0.8	0.18	Chinle	Oak Ridge 1 SE 7½′	
Shay Canyon	Grand	37°57′N	109°33′W	1.44	0.8	0.13	Morrison	Oak Ridge 1 NE 7½′	
Shay Ridge	Grand	37°50′N	109°33′W	3.51	1.44	0.54	Morrison	Oak Ridge 1 SE 7½′	
South Creek	Grand	37°49′N	109°27′W	1.28	0.96	0.24	Mancos sh	Verdure 2 SW 7½′	
South Peak	Grand	37°49′N	109°27′W	2.22	2.08	0.48	Morrison	Verdure 2 SW 7½′	
South Robertson Pasture	Grand	37°52′N	109°30′W	2.22	1.6	.43	Mancos sh	Verdure 2 NW 7½′ Verdure 2 SW 7½′ Elk 1 NE 7½′ and Elk 1 SE 7½′	
Spring Creek anticline	Grand	37°53′N	109°27′W	2.88	2.22	0.19	Morrison?	Verdure 2 NW 7½′	
Twin Peak	Grand	37°52′N	109°30′W	2.08	1.71	.43	Mancos sh	Elk Ridge 1 SE 7½′ and Verdure 2 SW 7½′	

COUNTRY State, Group, and Laccolith Name	County or District	Latitude	Longitude	Length km	Width km	Thickness km	Depth of Intrusion km or fm	Topographic Map	References and Remarks
Verdure	Grand	37°48′N	109°26′W	2.08	0.96	0.18	Dakota ss and Burro Canyon	Verdure 2 SW 7½′	
Viewpoint	Grand	37°49′N	109°30′W	0.32	0.16	0.13	Morrison	Oak Ridge 1 SE and Verdure 2 SW 7½′	

Henry Mountains group

COUNTRY State, Group, and Laccolith Name	County or District	Latitude	Longitude	Length km	Width km	Thickness km	Depth of Intrusion km or fm	Topographic Map	References and Remarks
Black Mesa	Garfield	37°54′N	110°37′W	1.6	1.6	0.16	1.4 Summerville	Mt. Pennell 1 NE 7½′	Gilbert (1877) Green (1879)
Buckhorn Ridge	Garfield	38°00′N	110°34′W	1.1	0.6	0.06	2.2 Chinle	Mt. Pennell 1 SE 7½′	Hunt (1938)
Bull Creek	Garfield	38°07′N	110°45′W	5.6	1.9	0.16	1.1 upper Morrison	Mt. Ellen 15′	Hunt (1946) Hunt and others (1953)
Bulldog Peak or Cass Creek Peak	Garfield	37°54′N	110°45′W	1.1	0.8	0.53	1.1 upper Morrison	Mt. Pennell 2 NW 7½′	Pollard (1969) Pollard and Johnson (1969)
Bullfrog Creek	Garfield	38°01′N	110°50′W	4.0	0.81	0.24	1.1 Tununk	Mt. Ellen 15′	Johnson (1970)
Bull Mountain	Garfield	38°08′N	110°44′W	2.9	1.6	0.81		Mt. Ellen 4 NW 7½′	Avakian (1970)
Butler Wash	Garfield	38°06′N	110°45′W	4.8	1.6	0.32	1.4 Summerville	Mt. Ellen 4 SW 7½′	Case and Joesting (1972) Pollard (1972)
Cedar Creek	Garfield	38°07′N	110°52′W	6.4	2.4	0.16	1.5 upper Entrada	Mt. Ellen 15′	Savage and Sowers (1972) Johnson and Pollard (1973)
Cedar Ridge	Garfield	38°07′N	110°52′W	3.5	1.1	0.24	1.1 upper Morrison	Mt. Ellen 15′	Pollard and Johnson (1973) Savage (1974)
Chaparral Hills	Garfield	37°54′N	110°39′W	4.8	1.9	0.24	1.1 Dakota	Mt. Pennell 1 NW 7½′	Corry (1976)
Copper Ridge	Garfield	38°02′N	110°46′W	5.6	3.5	0.40	1.1 Dakota	Mt. Ellen 15′	Kilinc (1979)
Corral Ridge or Point	Garfield	38°05′N	110°49′W	2.4	1.3	0.40	1.1 Tununk	Mt. Ellen 15′	Hunt (1980)
Coyote Creek	Garfield	37°58′N	110°46′W	1.9	1.6	0.32	1.1 upper Morrison	Mt. Pennell 2 NE 7½′	Affleck and Hunt (1980) Morton (1983)
Dark Canyon	Garfield	38°00′N	110°46′W	3.2	1.6	0.32	1.1 upper Morrison	Mt. Pennell 2 NE 7½′	Sullivan (1987) Jackson(1987)
Dugout Creek	Garfield	38°04′N	110°49′W	6.9	1.6	0.32	1.4 Summerville	Mt. Ellen 15′	Jackson and Pollard (1988)
Durfey Butte	Garfield	38°04′N	110°48′W	4.0	1.9	0.35	0.9 Ferron	Mt. Ellen 15′	
Granite Ridges	Garfield	38°05′N	110°46′W	3.5	2.4	0.48	1.1 upper Morrison	Mt. Ellen 15′	
The Horn	Garfield	37°59′N	110°47′W	4.0	2.4	0.40	0.9 Ferron	Mt. Pennell 2 NE 7½′	
Horseshoe Ridge	Garfield	38°08′N	110°47′W	4.2	3.2	0.48	1.1 Tununk	Mt. Ellen 15′	
NW of Mt. Ellen	Wayne and Garfield	38°08′N	110°48′W	3.2	1.9	0.64	1.1 upper Morrison	Mt. Ellen 15′	
Mt. Ellsworth	Garfield	37°44′N	110°37′W	4	4	1.8		Mt. Ellsworth and Mt. Hiller 15′	
Mt. Hiller	Garfield	37°53′N	110°42′W	6-7	6-7	2.5		Mt. Hiller 15′	
Mt. Holmes	Garfield	37°47′N	110°35′W	5-6	5-6	1.2		Mt. Hiller 15′	
Mt. Pennell	Garfield	37°57′N	110°47′W					Mt. Pennell 15′	
Nazer Canyon	Wayne and Garfield	38°09′N	110°47′W	4.2	3.2	0.48	1.1 Tununk	Mt. Ellen 15′	
North Sawtooth	Garfield	37°53′N	110°37′W	7.2	0.8	0.48	1.3 Morrison	Mt. Pennell 1 NE 7½′	
North Spur	Garfield	38°08′N	110°49′W	2.9	1.1	0.81	1.1 Tununk	Mt. Ellen 15′	
Pistol Ridge	Garfield	38°06′N	110°49′W					Mt. Ellen 15′	
Pulpit Arch	Garfield	37°53′N	110°35′W	5.1	1.6	0.16	1.1 Dakota	Mt. Pennell 1 NE 7½′	
Quaking Asp Creek	Garfield	37°55′N	110°43′W	3.2	0.8	0.16	1.1 Tununk	Mt. Pennell 1 NW 7½′	
Ragged Mountain	Garfield	38°01′N	110°45′W	1.6	1.6	0.48		Mt. Ellen 15′	
Sarvis Ridge	Garfield	38°04′N	110°48′W	5.6	0.8	0.16	1.1 Dakota	Mt. Ellen 15′	
Sawmill Basin	Garfield	38°07′N	110°46′W	3.2	2.1	0.16	0.9 Blue Gate	Mt. Ellen 15′	
Sawtooth Ridge	Garfield	37°52′N	110°37′W	7.2	0.8	0.48	1.3 Morrison	Mt. Hiller 1 NE 7½′	
Slate Creek	Garfield	38°03′N	110°47′W	3.2	1.6	0.48	1.1 upper Morrison	Mt. Ellen 15′	
South Creek	Garfield	38°04′N	110°50′W	4.8	1.9	0.40	0.9 Ferron	Mt. Ellen 15′	
Speck Canyon	Garfield	37°55′N	110°39′W	6.4	2.3	0.19	1.4 Summerville	Mt. Pennell 1 NW 7½′	
Specks Ridge	Garfield	37°55′N	110°40′W	2.4	0.8			Mt. Pennell 1 NW 7½′	
Stewart Ridge	Garfield	37°54′N	110°41′W	4.8	2.1	0.40	1.1 Tununk	Mt. Pennell 1 NW 7½′	
Table Mountain	Wayne	38°05′N	110°50′W	2.1	2.1	0.81	1.1 upper Morrison	Mt. Ellen 15′	
Theater Canyon	Garfield	37°47′N	110°34′W	0.8	0.8	0.06	2.0 Kayenta	Mt. Pennell 1 SE 7½′	
Trachyte Mesa	Garfield	37°57′N	110°35′W	0.8	0.75	0.03	Entrada	Mt. Pennell 2 NE 7½′	
Wickup Ridge	Garfield	38°06′N	110°48′W	2.4	1.1	0.64	1.1 upper Morrison	Mt. Ellen 15′	

High Plateaus group

COUNTRY State, Group, and Laccolith Name	County or District	Latitude	Longitude	Length km	Width km	Thickness km	Depth of Intrusion km or fm	Topographic Map	References and Remarks
Spry Intrusion	Garfield	38°03′N	112°24′W					Fremont Pass 7½′	Grant (1979) Grant and Anderson (1979)

COUNTRY State, Group, and Laccolith Name	County or District	Latitude	Longitude	Length km	Width km	Thickness km	Depth of Intrusion km or fm	Topographic Map	References and Remarks
La Sal Mountains group									
Beaver Creek	Grand	38°33′N	109°12′W	4.8	0.8	0.16		Mt. Waas 4 SW 7½′	Hill (1913)
Blue Lake and Horse Canyon	San Juan	38°28′N	109°14′W	1.6	1.6	0.32		Mt. Peale 1 NW and Mt. Peale 2 NE 7½′	Gould (1925a, 1925b), Gould (1926a, 1926b)
Brumley Creek	San Juan	38°28′N	109°16′W	2.4	0.8	0.16		Mt. Peale 2 NE 7½′	Hunt (1958)
Haystack Mountain	Grand and SanJuan	38°30′N	109°15′W	1.6	1.6	0.4		Mt. Peale 1 NW and Mt. Waas 4SW 7½′	Case and others (1963) Carter and Gualtieri (1965)
NE of North Mountain	Grand	38°35′N	109°14′W	5.6	3.2	0.64		Mt. Waas 3 SE and Mt. Waas 4 SW 7½′	Case and Joesting (1972) Hunt (1983)
NW of North Mountain	Grand	38°33′N	109°15′W	5.6	2.4	0.96		Mt. Waas 3 SE and Mt. Waas 4 SW 7½′	
SE of North Mountain	Grand	38°30′N	109°13′W	4.8	2.4	0.8		Mt. Waas 4 SW 7½′	
SW of North Mountain	Grand	38°33′N	109°18′W	4.45	3.2	0.8		Mt. Waas 3 SE 7½′	
Mt. Mellenthin	San Juan	38°28′N	109°14′W					Mt. Peale 1 NW 7½′	Feeder dikes visible
Mt. Peale	San Juan	38°27′N	109°14′W	3.2	2.0	0.64		Mt. Peale 1 NW 7½′	
Pack Creek	San Juan	38°25′N	109°17′W	3.2	0.8	0.32		Mt. Peale 2 NE 7½′	
SE of South Mountain	San Juan	38°20′N	109°12′W	4.8	1.6	0.32		Mt. Peale 1 SW 7½′	
Pine Valley group									
Granite Mountain	Iron	37°42′N	113°14′W			0.4		Cedar City NW 7½′	Leith and Harder (1908)
Iron Mountain	Iron	37°38′N	113°22′W					Iron Mountain 7½′	Young (1947)
Neck of the Desert	Iron	37°43′N	113°19′W					Desert Mound and Silver Peak 7½′	Mackin (1947a, 1947b) Cook (1952, 1953, 1957)
Paradise	Washington and Iron	37°33′N	113°27′W				0.6-0.9	Stoddard 7½′	Mackin (1960) Cook and Hardman (1967)
Pine Valley	Iron and Washington	37°27′N	113°23′W	32		0.9	>0.9 Claron >Tropic Navajo ss	New Harmony 15′	Mackin (1968) Spurney (1982)
Stoddard	Iron	37°35′N	113°25′W					Stoddard 7½′	
Three Peak	Iron	37°45′N	113°12′W					Three Peak 7½′	
House Range group									
Sawtooth Mountain	Millard	39°13′N	113°23′W	8				Notch Peak 15′	Butler and others (1920) Crawford and Buranek (1944)
Monument Valley group									
Navajo Mountain	San Juan	37°02′N	110°52′W	16		0.6		Navajo Mtn 15′	Baker (1936) Baker and others (1954)
Washington									
Cloudy Pass	Chelan and Snohomosh	48°13′N	120°55′W					Glacier Peak and Holden 15′	Carter (1969) Tabor and Crowder (1969)
Wyoming									
Absaroka group									
Clouds Home?	Park	44°15′N	109°44′W					Clouds Home Pk 7½′	Hague (1898)
Deer Creek	Park	44°10′N	109°39′W					Clouds Home Pk 7½′	Rouse (1933)
Dell Creek?	Park	44°03′N	109°51′W					Yellow Mtn 7½′	Love (1939)
Dell Creek stock	Park	44°04′N	109°52′W					Yellow Mtn 7½′	Parsons and others (1972)
Eagle Creek	Park	44°25′N	109°59′W					Eagle Creek 7½′	
Eagle Nest?	Park	44°22′N	109°58′W					Pinnacle Mtn 7½′	
Hurricane Mesa	Park	44°51′N	109°49′W					Pilot Peak 15′	
Ishawooa Mesa?	Park	44°11′N	109°41′W					Clouds Home Pk 7½′	
Needle Mountain?	Park	44°03′N	109°39′W					Needle Mtn 7½′	
Rampart Creek?	Park	44°18′N	109°46′W					Sheep Mesa 7½′	
Stinkingwater Peak	Park	44°38′N	109°45′W					Sunlight Peak 15′	
E. of Wapita River?	Park	44°21′N	109°41′W					Sheep Mesa 7½′	
Washakie Needles?	Fremont	43°47′N	109°12′W					Twin Peaks 7½′	
Windy Mountain?	Park	44°48′N	109°34′W					Beartooth Butte 15′	
Yellow Mountain?	Park	44°01′N	109°45′W					Yellow Mtn 7½′	
Gallatin Range group									
Bacon Rind anticline?	Gallatin and Yellowstone N.P.	44°56′N	111°05′W					Tepee Creek 15′	Hague (1896) Peale (1896)
Bunsen Peak?	Yellowstone	44°56′N	110°42′W					Mammoth 15′	Iddings and Weed (1899)

COUNTRY State, Group, and Laccolith Name	County or District	Latitude	Longitude	Length km	Width km	Thickness km	Depth of Intrusion km or fm	Topographic Map	References and Remarks
Dome Mountain	Yellowstone	44°50′N	110°51′W					Mt. Holmes 15′	Hague and others (1899)
Electric Peak?	Yellowstone	45°01′N	110°49′W					Miner 15′	Witkind (1964b, 1969)
Gallatin River	Yellowstone	45°00′N	111°04′W	6.4	3.2	0.61		Tepee Creek 15′	Ruppel (1972)
Gray Peak-Snowshoe	Yellowstone	44°57′N	110°53′W					Mt. Holmes 15′	
Indian Creek	Yellowstone	44°52′N	110°52′W					Mt. Holmes 15′	
N. of Joseph Peak	Yellowstone	44°43′N	110°54′W					Mammoth 15′	
Little Quadrant Mtn	Yellowstone	44°58′N	110°51′W					Mt. Holmes 15′	
Mt. Holmes stock?	Yellowstone	44°49′N	110°53′W					Mammoth and Mt. Holmes 15′	
Trilobite Point	Yellowstone	44°49′N	110°51′W					Mt. Holmes 15′	

UNITED SOVIET SOCIALIST REPUBLIC

Azerbaijan

Jeleznaia		?	?						Milanovsky and
Zmeievaia		?	?						Koronovsky (1966)
Pyatigorsk		?	?						Gerasimov (1974)

Laccoliths are reported in the Pyatigorsk area but no details are available as to exact locations.

Levskiy and Rubley (1978)
Gurbanov and Favorskya (1978)

Caucasian Mountains group

Beshtau Gora		44°07′N	43°01′E						Derweiss (1903)
Gora Byk		44°13′N	42°59′E						Guerassimov (1937)
Razvalka Gora		44°10′N	43°01′W						

Georgia

Merisi intrusion		41°33′N?	41°43′E?						Kebuladze and Tatishvili (1969)

Kazakhstan

Akzhaylautas		47°28′N	82°00′E						Sviridenko (1976)
Sibiny		49°18′N?	83°00′E?						Kravchenko and Beskin (1979)

Kirgiz

Babaytag Massif		41°20′N?	70°55′E?						Kantsel and others (1972)
Samgarskiy		40°35′N?	69°35′E?						Babakhodzhayev and Tadzhibayev (1972)

Russian Soviet Federated Socialist Republic (Siberia)

Barguzin complex		53°50′N?	109°30′E?						Altukhov (1974)
Murunsk Massif		60°05′N?	120°00′E?						Vashchilov (1963)
Tegir-Tyz Massif		54°45′N?	88°00′E?						Lazebnik and Lazebnik (1979)
Yano-Kolymskoy		61°30′N?	150°00′E?						Bazhenov and others (1981)
Synnyr		56°45′N?	111°00′E?						Zhidkov (1978)

Tadzhikistan

Pamir group

Lyangar		37°02′N?	72°42′E?						Khasanov (1968)

Ural Mountains group

Kempirary Massif		55°00′N?	60°00′E?						Segalovich (1971)

APPENDIX C

NOTATION USED FOR LACCOLITH MODELS

a Radius of punched laccolith.

a_i Radius of the ith level of a Christmas-tree laccolith.

a_n Radius of deepest sill of Christmas-tree laccolith.

$\bar{D}$ Average crack width between $-L < z < L$.

$D(z)$ Lateral displacement of the feeder dike, or crack wall at any point z, where $-L < z < L$, and $D(z) = 0$ for $|z| > L_c$.

e Volumetric strain.

$E_t(\epsilon)$ Tangent modulus at a given strain, ϵ.

F Forces acting upward on the roof of an intrusion. Equivalent to the total magma driving pressure (P_t) times the area over which the magma pressure is applied, $P_t \pi r^2$.

f_s Spatial frequency. $f_s = 1/\lambda$. Units are cycles per unit length.

f_N The Nyquist frequency. Defined by the relation $f_N = 1/2\Delta T$ in the time domain and $f_N = 1/2\Delta\lambda$ in the space domain, where ΔT is the sample interval in seconds in the time domain, and $\Delta\lambda$ is the sample interval in units of length in the space domain. The Nyquist frequency is the highest frequency that can be reproduced from a discrete sampling interval of a continuous function. The Nyquist sampling theorem requires at least two samples per cycle of the highest frequency contained in the data to be reproduced by discrete sampling. If the continuous function contains higher frequencies than the Nyquist frequency then aliasing will occur due to inadequate sampling. Aliasing is the introduction of spurious, low-frequency anomalies into the waveform reproduced from the discrete samples. For a frequency $Y > f_N$, an aliased frequency $f_N - Y$ will be introduced into the function by aliasing, or folding.

$G(e)$ Shear modulus of the overburden, or country rock. G is a nonlinear function of the volumetric strain, e, for most rocks.

g Acceleration due to gravity.

h Height of the observer above the bottom of the dislocation crack.

J_2 Second stress invariant.

$K(e)$ Bulk modulus of the overburden, or country rock. K is a nonlinear function of the volumetric strain, e, for most rocks.

k Yield limit in simple shear.

k' Yield stress required to cause flow in a magma that behaves as a Bingham substance.

L The half length of the feeder dike, or dislocation crack, where the crack extends from $-L < z < L$.

L_c Critical crack length. For the region $|L| > L_c$ then $D(z) = 0$, that is, the crack is closed for $|L| > L_c$.

P_d Magma driving pressure. Equal to total magma pressure minus lithostatic pressure, $P_t - P_l$. For the feeder dike, or dislocation crack, to exist the relation $P_d > 0$ must hold.

P_l The lithostatic pressure. Generally taken as $\bar{\rho}_2 gz$ but may require correction for hydrostatic pressure to obtain effective lithostatic pressure. Effective pressure has not been calculated unless specifically so stated.

P_t The total magma pressure. Equal to the lithostatic pressure, P_l, plus the magma driving pressure, P_d.

Q Forces acting downward on the roof of the intrusion. In the simplest case of uniform cylindrical loading on the base of the plate, Q is equivalent to the shear strength of the plate times the area of a cylindrical surface extending from the plane of intrusion to the surface around the periphery, $2\pi r k z_l$, plus the unit weight of the plate times the plate thickness and the area of the intrusion, $\pi \bar{\rho} r^2 gz_l$.

r Radial axis parallel to the earth's surface. r is taken to be zero at the center of the intrusion. In the included models the intrusion is taken to be axisymmetric about the $r = 0$ axis. r, θ, and z axes are orthogonal.

r_i Radius of the ith sill in a Christmas-tree laccolith.

T Period of a continuous function in the time domain. Usually the function is represented by a Fourier series. Units are seconds.

ΔT Sample interval in seconds.

u Displacement in the x direction, or θ direction in cylindrical coordinates.

V Total volume of magma per unit width of crack length within the feeder dike. The feeder dike, or crack, is considered to extend to plus or minus infinity in the vertical plane of the crack.

v Displacement in the y direction, or r direction in cylindrical coordinates.

w Displacement in the z (vertical) direction.

w_0 The limital thickness of a laccolith, that is, the maximum thickness a laccolith can attain if the roof remains in place during growth and the floor is not depressed by the loading.

x Axis parallel to the earth's surface. For elliptical intrusions, x is taken as the minor axis. The x, y, and z axes are the orthogonal Cartesian coordinates.

y Axis parallel to the earth's surface. For elliptical intrusions y is taken as the major axis.

z_e Effective thickness of a layered, elastic overburden.

$z(h)$ The value of z at a distance h above the bottom of the crack.

z_i	Depth of intrusion of the ith sill of a Christmas-tree laccolith. There are 1 to n subsidiary sills and n may exceed 100.
z_l	Depth of intrusion for the sill that eventually forms a punched laccolith.
z_n	Depth of intrusion of the deepest sill in a Christmas-tree laccolith.
α	$1 - \nu$
ϵ	Two-dimensional strain.
θ	Axis parallel to the earth's surface in cylindrical coordinates. Measured counterclockwise relative to the x-axis for conversion to Cartesian coordinates.
λ	Wavelength of a function in the space domain. Units for gravity surveys are commonly kilometers. The inverse, $1/\lambda$, is called the spatial frequency. Spatial frequency units are then cycles per kilometer.
$\Delta\lambda$	Sample interval in the same length units as the spatial wavelength.
ν	Poisson's ratio.
$\nu(e)$	Poisson's ratio is, in general, a function of the volumetric strain, e.
ρ	Density of the country rock at a given depth.
$\bar{\rho}$	Weighted mean density of the country rock. The weighting function is the thickness of each layer of different density.
$\bar{\rho}_1$	Weighted mean density of the country rock evaluated at depth $2L_c + z_l$ for a punched laccolith.
$\bar{\rho}_z$	Weighted mean density of the country rock evaluated at depth z.
ρ'	Magma density. Assumed to remain constant with depth (i.e., incompressible). May also represent a magma that is a crystalline mush.
$\Delta\rho$	Density contrast. Density of body perturbing the gravity field minus country rock density.
σ_i	Normal stress in direction of axis indicated by subscript. Compressive stresses are taken to be positive.
τ_{ij}	Shear stress along axes designated by subscripts. Shear stresses are taken to be positive in the usual Timoshenko convention.
τ_0	Cohesive strength of the country rock measured normal to the plane of the intrusion unless otherwise stated.

REFERENCES CITED

Abel, K. D., Himmelberg, G. R., and Ford, A. B., 1979, Petrologic studies of Dufek intrusion; Plagioclase variation: Antarctic Journal of the United States, v. 14, no. 5, p. 6–8.

Adam, K. D., 1979, Changing views in the earth sciences illustrated by the example of the Steinheim Basin, Wurttemberg: Naturhistorisches Museum in Wien, Annalen, v. 83, p. 13–23.

Affleck, J., and Hunt, C. B., 1980, Magnetic anomalies and structural geology of stocks and laccoliths in the Henry Mountains, Utah, in Picard, M. D., ed., Henry Mountains symposium: Utah Geological Association Publication no. 8, p. 107–112.

Albritton, C. C., Jr., and Smith, J. F., 1965, Geology of the Sierra Blanca area, Hudspeth County, Texas: U.S. Geological Survey Professional Paper 479, 131 p.

Aldrich, M. J., Jr., 1974, Structural development of the Hanover-Fierro pluton, southwestern New Mexico: Geological Society of America Bulletin, v. 85, no. 6, p. 963–968.

Allan, J. A., 1914, Geology of Field map-area, British Columbia and Alberta: Canada Geological Survey Memoir 55, 312 p., scale 1:126,720.

Alling, H. L., 1919, Some problems of the Adirondack pre-Cambrian: American Journal of Science, 4th Series, v. 48, p. 47–68.

Altukov, Ye. N., 1974, New data on the geologic structure of the Barguzin-Vitim interfluve, western Transbaikal: Doklady of the Academy of Sciences of the USSR, Earth Science Sections 215, no. 1–6, p. 71–73.

Anderson, E. M., 1938, The dynamics of sheet intrusion: Royal Society of Edinburgh Proceedings, v. 58, p. 242–251.

—— , 1951, Dynamics of faulting and dyke formation with application to Britain (second edition): Edinburgh, Oliver and Boyd Limited, 206 p.

Anna, L., 1973, Geology of the Kirk Hill area, Meade County, South Dakota [M.S. thesis]: Rapid City, South Dakota School of Mines and Technology, 47 p.

Aramiki, S., 1968, Segregation vein in the Uwekahuna laccolith, Kilauea caldera, Hawaii: University of Tokyo, Bulletin of the Earthquake Research Institute, v. 46, Part 1, p. 155–160.

Armstrong, R. L., 1969, K-Ar dating of laccolithic centers of the Colorado plateau and vicinity: Geological Society of America Bulletin, v. 80, p. 2081–2086.

Aubele, J. C., and Crumpler, L. S., 1983, Geology of the central and eastern parts of the Springerville–Show Low volcanic field, Arizona: Geological Society of America Abstracts with Programs, v. 15, no. 5, p. 303.

Autran, A., and 11 others, 1979, The Central Massif: Revue des Sciences Naturelles D'Auvergne, v. 45, no. 1–4, p. 010.29–010.30.

Avakian, R. W., 1970, Determination of the subsurface geology of Mt. Hillers, Utah, by use of geophysical techniques; Gravity and magnetics: Stanford, Stanford University Technical Report, 97 p.

Babakhodzhayev, S. M., and Tadzhibayev, G. T., 1972, Geologic and petrographic facial characteristics of the Samgarskiy volcanic-plutonic complex in eastern Karamazar: Akademiya Tadzhikskoy Izvestiya Otdeleniye; Fizika Matematicheskikh, Geologicheskikh, Khimicheskikh Nauk, no. 2, p. 93–102.

Babin, C., Didier, J., and Jonin, M., 1968, Aicrogranite laccolith in the Brest roads; Longue Island: France, Bureau de Recherches Geologiques et Minieres, Bulletin, Series 2, sec. 1, no. 3, p. 1–8.

Badham, J.P.N., 1981, Petrochemistry of late Aphebian (differs from 1.8 Ga.) calc-alkaline diorites from the East Arm of Great Slave Lake, Northwest Territories, Canada: Canadian Journal of Earth Sciences v. 18, no. 6, p. 1018–1028.

Badham, J.P.N., and Nash, J. T., 1978, Magnetite-apatite-amphibole-uranium and silver-arsenide mineralizations in lower Proterozoic igneous rocks, East Arm, Great Slave Lake, Canada, in Nash, J. T., ed., Uranium geology in resource evaluation and exploration: Economic Geology, v. 73, no. 8, p. 1474–1491.

Bagstad, D. P., 1981, Structural analysis of folding of Paleozoic sequence, Solita-

rio uplift, Trans-Pecos Texas [M.S. thesis]: Lubbock, Texas Tech University, 42 p.

Bailey, E. B., 1912, The new mountain of the year 1910, Usu-san, Japan: Geological Magazine, v. 9, p. 248–252.

—— , 1926, Domes in Scotland and South Africa, Arran and Vredefort: Geological Magazine, v. 63, p. 487–488.

Baker, A. A., 1936, Geology of the Monument Valley–Navajo Mountain region, San Juan County, Utah: U.S. Geological Survey Bulletin 865, 106 p.

Baker, A. A., Clark, L. W., Kelley, L. A., Snow, L. G., and Larsen, R. M., 1954, Preliminary map showing geologic structure of the Monument Valley–Navajo Mountain region, San Juan County, Utah: U.S. Geological Survey Oil and Gas Investigations Map OM-168, scale 1:126,720.

Baldwin, W. M., Brown, R. D., Jr., Gair, J. E., and Pease, M. H., Jr., 1955, Geology of the Sheridan and McMinnville quadrangles, Oregon: U.S. Geological Survey Oil and Gas Investigations Map OM-155, scale 1:26,270.

Balsley, J. R., Gilbert, P. P., Mangan, G. B., and 8 others, 1957a, Aeromagnetic map of the Laredo Quadrangle, Bearpaw Mountains, Montana: U.S. Geological Survey Geophysical Investigations Map GP-150, scale 1:31,680.

—— , 1957b, Aeromagnetic map of the Shambo Quadrangle, Bearpaw Mountains, Montana: U.S. Geological Survey Geophysical Investigations Map GP-151, scale 1:31,680.

—— , 1957c, Aeromagnetic map of the Centennial Mountain Quadrange, Bearpaw Mountains, Montana: U.S. Geological Survey Geophysical Investigations Map GP-152, scale 1:31,680.

—— , 1957d, Aeromagnetic map of the Warrick Quadrangle, Bearpaw Mountains, Montana: U.S. Geological Survey Geophysical Investigations Map GP-153, scale 1:31,680.

Baltzer, A., 1904, Die granitschen lakkolithenartiegen intrusion massen des Aarmassivs: International Geological Congress, Vienna, 9th session, Compte Rendu 2, p. 787–789.

Barker, D. S., Long, L. E., Hoops, G. K., and Hodges, F. N., 1977, Petrology and Rb-Sr isotope geochemistry of intrusions in the Diablo Plateau, northern Trans-Pecos magmatic province, Texas and New Mexico: Geological Society of America Bulletin, v. 88, no. 10, p. 1437–1446.

Barksdale, J. D., 1937, The Shonkin Sag laccolith, Montana: American Journal of Science, 5th Series, v. 33, no. 197, p. 321–359.

—— , 1950, Shonkin Sag laccolith revisited (Montana) [abs.]: Geological Society of America Bulletin, v. 61, no. 12, part 2, p. 1442 (also see American Mineralogist, v. 36, no. 3–4, p. 310, 1951).

—— , 1952, Pegmatite layer in the Shonkin Sag laccolith, Montana: American Journal of Science, v. 250, no. 10, p. 705–720.

Bass, N. W., Eby, J. B., and Campbell, M. R., 1955, Geology and mineral fuels of parts of Routt and Moffatt Counties, Colorado: U.S. Geological Survey Bulletin 1027-D, p. 143–250.

Bassett, W. A., 1961, Potassium-argon age of Devils Tower, Wyoming: Science, v. 134, no. 3487, p. 1373.

Bateman, J. D., and Harrison, J. M., 1944, Mikanagan Lake, Manitoba: Canada Geological Survey Paper 44-22, 6 p. (reprinted in Precambrian, v. 17, no. 10, p. 4–7, 1944).

Bates, R. L., and Jackson, J. A., eds., 1980, Glossary of geology (second edition): Falls Church, American Geological Institute, 751 p.

Bates, T. E., 1980, The origin, distribution, and geological setting of copper and nickel sulphides in the Riwaka complex, northwest Nelson, New Zealand: New Zealand Conference no. 9, Australian Institute of Mining and Metallurgy, p. 35–51.

Bayley, W. S., 1894, The basic massive rocks of the Lake Superior region IV, A, B: Journal of Geology, v. 2, p. 814–825.

—— , 1895, The basic massive rocks of the Lake Superior region IV, C: Journal of Geology, v. 3, p. 1–20.

Bazhenov, A. I., Myshko, Z. A., and Polvektova, T. I., 1981, Magnetic properties of granitoids and distribution of ferromagnetic minerals as petrogenetic criteria (as exemplified in the Tegir-Tyz Massif): International Geology Review,

v. 25, no. 1, p. 39–44.

Beall, J. J., 1972, Pseudo-rhythmic layering in the Square Butte alkali-gabbro laccolith: American Mineralogist, v. 57, no. 7, 8, p. 1294–1302.

——, 1973, Mechanics of intrusion and petrochemical evolution of the Adel Mountain volcanics [Ph.D. thesis]: Missoula, University of Montana, 115 p.

——, 1977, Intrusive geology of the Adel Mountain volcanics, northern Big Belt Range, Montana: Geological Society of America Abstracts with Programs, v. 9, no. 6, p. 707–708.

Beck, J. A., Jr., 1976, Geology of the Lexington Hill-Pillar Peak area, Lawrence County, South Dakota [M.S. thesis]: Rapid City, South Dakota School of Mines and Technology, 102 p.

Beck, J. A., and Lisenbee, A. L., 1981, Sequential development of a primary and secondary laccolith southeast of Deadwood, South Dakota: Geological Society of America Abstract with Programs, v. 13, no. 4, p. 190.

Behrendt, J. C., Henderson, J. R., Meister, L., and Rambo, W. L., 1974, Geophysical investigations of the Pensacola Mountains and adjacent glacierized areas of Antarctica: U.S. Geological Survey Professional Paper 844, 27 p.

Behrendt, J. C., Drewry, D. J., Jankowski, E., and England, A. W., 1979, Revision of known area of Dufek intrusion: Antarctic Journal of the United States, v. 14, no. 5, p. 6.

Bergendahl, M. H., and Koschmann, A. H., 1971, Ore deposits of the Kokomo-Tenmile district, Colorado: U.S. Geological Survey Professional Paper 652, 53 p.

Bermingham, P. M., Fairhead, J. D., and Stuart, G. W., 1983, Gravity study of the Central African Rift system; A model of continental disruption; Part 2, The Darfur domal uplift and associated Cenozoic volcanism: Tectonophysics, v. 94, no. 1–4, p. 205–222.

Bevier, M. L., and Klebanow, R. S., 1984, Geology and petrography of Crested Butte laccolith: Geological Society of America Abstracts with Programs, v. 16, no. 4, p. 215.

Bhattacharya, B. K., and Chan, K. C., 1977, Reduction of magnetic and gravity data on an arbitrary surface acquired in a region of high topographic relief: Geophysics, v. 42, no. 7, p. 1411–1430.

Bhattacharya, B. K., and Navolio, M. E., 1975, Digital convolution for computing gravity and magnetic anomalies due to arbitrary bodies: Geophysics, v. 40, no. 6, p. 981–992.

Billings, M. P., 1972, Structural geology (third edition): Englewood Cliffs, Prentice-Hall, Incorporated, 606 p.

Bixel, F., 1973, Structure of the volcanic complex of the Ossau, Basses-Pyrenees: France, Bureau de Recherches Geologiques et Minieres, Bulletin Serie 2, Sect. 1, no. 3, p. 165–178.

Blake, D. H., 1980, Volcanic rocks of the Paleohelikian Dubawnt Group in the Baker Lake Angikuni Lake area, District of Keewatin, Northwest Territories: Canada, Geological Survey Bulletin 309, 39 p.

Blake, J. F., 1898, The laccolites of Cutch and their relations to the other igneous masses of the district: Geological Society of London, Quarterly Journal, Abstract, v. 54, p. 12.

Blixt, J. E., 1933, Geology and gold deposits of the North Moccasin Mountains, Fergus County, Montana: Montana Bureau of Mines and Geology, Memoir 8, 25 p.

Bonini, W. E., Biehler, S., Clermont, K. M., Kelly, W. N., Jr., and Vreeland, J. H., 1967, The gravitational field and the shape of intrusive bodies: Department of Civil and Geological Engineering, Princeton University, Geological Engineering Report 67-3, 26 p.

Books, K. G., 1962, Remanent magnetism as a contributor to some aeromagnetic anomalies: Geophysics, v. 27, no. 3, p. 359–375.

Bothner, W. A., 1974, Gravity study of the Exeter pluton, southeastern New Hampshire: Geological Society of America Bulletin, v. 85, no. 1, p. 51–56.

Bradley, J., 1965, Intrusion of major dolerite sills: Royal Society of New Zealand, Transactions, Geology, v. 3, no. 4, p. 27–55.

Brammall, A., and Harwood, F., 1932, The Dartmoor granites; Their genetic relationships: Quarterly Journal of the Geological Society of London, v. 88, p. 171–237.

Brawner, C. O., 1974, Rock mechanics in open pit mining, in Advances in Rock

Mechanics: International Society for Rock Mechanics, Congress, Proceedings, no. 3, v. 1, part A, p. 755–815.

Breed, W. J., 1971, Is Preston Mesa a laccolith?: Plateau, v. 44, no. 2, p. 78–80.

Briden, J. C., Clark, R. A., and Fairhead, J. D., 1982, Gravity and magnetic studies in the Channel Islands: Journal of the Geological Society of London, v. 139, p. 35–48.

Bridgwater, D., Sutton, J., and Watterson, J., 1974, Crustal downfolding associated with igneous activity: Tectonophysics, v. 21, no. 1–2, p. 57–77.

Brocher, T. M., 1978, Analysis of a magma body reflection of Abo Pass, New Mexico, COCORP data [abs.]: EOS, Transactions, American Geophysical Union, v. 59, no. 4, p. 390.

Brögger, W. C., 1895, Der Mechanismns der Eruption der Tiefengesteine: Die Eruptivgestine des Kristianiagebietes, v. 2, p. 118–119.

Brouwer, H. A., 1910, Oorsprong en Samenstelling der Transvaalsche Nepheliensyenietien's: Gravenhage, Drukkerij Mouton and Company, 180 p.

Brown, B. W., 1952, A study of the southern Bear Lodge Mountains intrusive [M.S. thesis]: Lincoln, University of Nebraska, 63 p.

——, 1956, A study of the Northern Black Hills Tertiary petrogenic province with notes on the geomorphology involved [Ph.D. thesis]: Lincoln, University of Nebraska, Dissertation Abstracts, v. 14, no. 8, p. 1198–1199 (1954).

Brown, B. W., and Lugn, A. L., 1952, Study of the Bear Lodge Mountains intrusive, Wyoming [abs.]: Geological Society of America Bulletin, v. 63, no. 12, part 2, p. 1378 (also see Nebraska Academy of Science Proceedings, 63rd Annual Meeting, p. 11, 1953).

Brown, I. A., 1930, The geology of the south coast of New South Wales: Linnean Society of New South Wales, Proceedings, v. 55, p. 637–698.

Bryant, B., Schmidt, R. G., and Pecora, W. T., 1960, Geology of the Maddux Quadrangle, Bearpaw Mountains, Blaine County, Montana: U.S. Geological Survey Bulletin 1081-C, p. 91–116.

Buchmann, J. P., 1960, Exploration of a geophysical anomaly at Trompsburg, Orange Free State, South Africa: Transactions of the Geological Society of South Africa, v. 63, p. 1–10.

Buda, G., 1966, Statistiche Verteilung und qualitative Kennzeichnung der Feldspate im Andesit-Lakkolit des Csodi-Berges: Budapest, University of Science Annals, Geology Section, v. 9, p. 123–131 (1965).

Buddington, A. F., 1929, Granite phacoliths and their contact zones in the northwest Adirondacks: New York State Museum Bulletin 281, p. 51–107.

——, 1939, Adirondack igneous rocks and their metamorphism: Geological Society of America Memoir 7, 354 p.

——, 1948, Origin of the granitic rocks of the northwest Adirondacks: Geological Society of America Memoir 28, p. 21–43.

Buddington, A. F., and Hess, H. H., 1937, Layered peridotite laccoliths in the Trout River area, Newfoundland (a discussion): American Journal of Science, 5th Series, v. 33, no. 197, p. 380–388.

Burbank, W. S., 1930, Revision of geologic structure and stratigraphy in the Ouray district of Colorado, and its bearing on ore deposition: Colorado Scientific Society Proceedings, v. 12, no. 6, p. 151–232.

Burckhardt, C., 1906, Geologie do la Sierra de Mazapil et Santa Rosa: International Geological Congress, 10th session, Guide des Excursion, no. 26 (Excursion du Nord), 14 plates, 40 p.

Burton, J. C., and Jackson, D. H., 1979, Contact metamorphism by convective heat transfer at Wind Mountain, New Mexico: Neues Jahrbuch fur Mineralogie, Monatshelfte, v. 9, p. 408–420.

Butler, B. S., Loughlin, G. F., Heikes, V. D., and others, 1920, The ore deposits of Utah: U.S. Geological Survey Professional Paper 111, 672 p.

Cady, G. H., and Dapples, E. E., 1937, Laccoliths of the Crested Butte anthracite district, Colorado [abs.]: Geological Society of America Proceedings 1936, June 1937, p. 67.

Cannon, R. S., 1937, Geology of the Piseio Lake Quadrangle: New York State Museum Bulletin 312, 107 p.

Carter, F. W., 1969, The Cloudy Pass epizonal batholith and associated subvolcanic rocks: Geological Society of America Special Paper 116, 54 p.

Carter, W. D., and Gualtieri, J. L., 1965, Geology and uranium-vanadium deposits of the La Sal Quadrangle, San Juan County, Utah, and Montrose County,

Colorado: U.S. Geological Survey Professional Paper 508, 82 p.

Case, J. E., and Joesting, H. R., 1972, Regional geophysical investigations in the Colorado Plateau: U.S. Geological Survey Professional Paper 736, 31 p.

Case, J. E., Joesting, H. R., and Byerly, P. E., 1963, Regional geophysical investigations in the La Sal Mountains area, Utah and Colorado: U.S. Geological Survey Professional Paper 316-F, p. 91–116.

Chadwick, G. H., 1942, Laccoliths of Frenchman's Bay, Maine [abs.]: Geological Society of America Bulletin, v. 53, no. 12, part 2, p. 1797.

—— , 1944, The geology of Mount Desert Island (Acadia National Park), Maine: New York Academy of Sciences Transactions, Series 2, v. 6, no. 6, p. 171–178.

Chapman, C. A., 1939, Geology of the Mascoma Quadrangle, New Hampshire: Geological Society of America Bulletin, v. 50, no. 1, p. 127–180.

—— , 1942, Intrusive domes of the Claremont-Newport area, New Hampshire: Geological Society of America Bulletin, v. 53, no. 6, p. 889–915.

—— , 1969, Oriented inclusions in granite, further evidence for floored magma chambers: American Journal of Science, v. 267, no. 8, p. 988–998.

Chattopadhyay, B., and Saha, A. K., 1974, The Neropahar pluton in eastern India; A model of Precambrian diapiric intrusion: Neues Jahrbuch fur Mineralogie, Abhandlungen, v. 121, no. 2, p. 103–126.

Chenevoy, M., 1967, Micro-crystalline rocks of the Veranne area, Loire (Saint Etienne Quadrangle): France, Service del la Carte Geologique, Bulletin, v. 61, no. 278, p. 141–152.

Christiansen, H., and Stephenson, M., 1983, MOVIE.BYU: Civil Engineering Department, Brigham Young University, 368 Clyde Building, Provo, Utah 84602.

Clark, T. H., 1972, Stratigraphy and structure of the St. Lawrence lowland of Quebec: International Geological Congressional Guidebook, no. 24, part C52, 82 p.

Cook, E. F., 1952, Geology of the Pine Valley Mountains; A preliminary note, *in* Guidebook to the Geology of Utah: Utah Geological Society, no. 7, p. 92–100.

—— , 1953, Pine Valley laccolith, Washington County, Utah [abs.]: Geological Society of America Bulletin, v. 64, no. 12, part 2, p. 1543.

—— , 1957, Geology of the Pine Valley Mountains, Utah: Utah Geological and Mineralogical Survey Bulletin, v. 58, 111 p.

Cook, K. L., and Hardman, E., 1967, Regional gravity survey of the Hurricane fault area and Iron Springs district: Geological Society of America Bulletin, v. 78, no. 9, p. 1063–1076.

Cooke, H. C., 1930, The compound laccolith of Lake Dufault, Quebec: The Royal Society of Canada Transactions, Series 3, v. 24, p. 89–98.

Cooke, H. C., James, W. F., and Mawdsley, J. B., 1931, Geology and ore deposits of Rouyn-Harricanaw region, Quebec: Canada Geological Survey Memoir 166, 314 p.

Cornu, F., 1906, Petrographische untersuchung einiger enallogener Einschlusse aus den Trachyten der Euganeen: Beitraege zur Paleontologie und Geologie Oesterreich-Ungarns und des Orients, p. 45.

Corry, C. E., 1972, The origin of the Solitario, Trans-Pecos, Texas [M.S. thesis]: Salt Lake City, University of Utah, 151 p.

—— , 1976, The emplacement and growth of laccoliths [Ph.D. thesis]: College Station, Texas A&M University, 184 p.

Corry, C. E., Herrin, E., McDowell, F., and Phillips, K., 1988, Geology of the Solitario, Trans-Pecos Texas: The University of Texas at Austin, Bureau of Economic Geology, Report of Investigations, in press.

Couturie, J., Vachette-Caen, M., and Vialette, Y., 1979, Namurian age of a gravity-differentiated granitic laccolith; The Margeride Granite; French Central Massif: Academy of Science (Paris), Comptes Rendu, Serie D289, no. 5, p. 449–452.

Crawford, A. L., and Buranek, A. M., 1944, Amazon stone, a new variety of feldspar for Utah, with notes on the laccolithic character of the House Range intrusive [abs.]: Utah Academy of Science Proceedings, 1941–1943, v. 19–20, p. 125–127.

Crawford, R. D., 1913, Geology and ore deposits of the Monarch and Tomichi districts, Colorado: Colorado Geological Survey, Bulletin 4, 101 p.

Cross, W., 1894, The laccolithic mountain groups of Colorado, Utah, and Arizona: U.S. Geological Survey, 14th Annual Report, Part IId, p. 165–241 (also see reviews by Pirsson, L. V., 1895, American Journal of Science, Series 3, v. 50, p. 74; and Iddings, J. P., 1895, Journal of Geology, v. 3, p. 853–856).

—— , 1898, Igneous rocks of the Telluride district, Colorado: Colorado Scientific Society Proceedings, v. 5, p. 225–234.

Cross, W., and Hole, A. D., 1910, Description of the Engineer Mountain Quadrangle, Colorado: U.S. Geological Survey Geological Atlas Folio 171, 101 p.

Cross, W., and Purington, C. W., 1899, Description of the Telluride Quadrangle, Colorado: U.S. Geological Survey Geological Atlas Folio 57, 19 p.

Cross, W., and Ransome, F. L., 1905, Description of the Rico Quadrangle, Colorado: U.S. Geological Survey Geological Atlas Folio 130, 20 p.

Cross, W., and Spencer, A. C., 1900, Geology of the Rico Mountains, Colorado: U.S. Geological Survey, 21st Annual Report, part 2, p. 7.

Cross, W., Spencer, A. C., and Purington, C. W., 1899, Description of the La Plata Quadrangle, Colorado: U.S. Geological Survey Geological Atlas Folio 60, 14 p.

Cross, W., Howe, E., and Ransome, F. L., 1905, Description of the Silverton Quadrangle, Colorado: U.S. Geological Survey Geological Atlas Folio 120, 18 p.

Cunningham, C. C., Naeser, C. W., and Marvin, R. F., 1977, New ages for intrusive rocks in the Colorado Mineral Belt: U.S. Geological Survey Open-File Report 77-573, 7 p.

Currie, K. L., 1969, Geological notes on the Carswell circular structure: Canada, Geological Survey, Paper 67-32, 60 p.

Cwojdzinski, S., 1975, Structural evolution of the Klodzko–Zloty Stok granitoid intrusion in relation to the tectonics of its envelope: Kwartalnik Geolgiczny, v. 19, no. 2, p. 472–473.

Daly, R. A., 1914, Igneous rocks and their origin: New York, McGraw-Hill, p. 63–78.

—— , 1933, Igneous rocks and the depths of the earth: New York, McGraw-Hill, 598 p.

Daly, R. A., Manger, G. E., and Clark, S. P., Jr., 1966, Density of rocks, *in* Clark, S. P., Jr., ed., Handbook of physical constants: Geological Society of America Memoir 97, p. 19–26.

Dapples, E. C., 1940, The distribution of heavy accessory minerals in a laccolith; Colorado: American Journal of Science, v. 238, no. 6, p. 439–450.

Darracott, B. W., 1974, A Precambrian laccolithic intrusion at Mara, northern Tanzania, revealed by gravity and magnetic surveys: Transactions of the Geological Society of South Africa, v. 77, no. 2, p. 79–84.

Darton, N. H., 1905, Sundance, Wyoming–South Dakota geologic folio: U.S. Geological Survey Geological Atlas Folio GF-127, 12 p.

—— , 1909, Geology and water resources of the northern portion of the Black Hills and adjoining regions in South Dakota and Wyoming: U.S. Geological Survey Professional Paper 65, 105 p.

Darton, N. H., and Paige, S., 1925, Description of the central Black Hills, South Dakota: U.S. Geological Survey Geological Atlas Folio 219, 34 p.

Daugherty, F. W., 1963a, Geology of Pico Etereo area, Municipio de Acuna, Coahuila, Mexico [abs.]: Geological Society of America Special Paper 73, p. 136.

—— , 1963b, La Cueva intrusive complex and dome, northern Coahuila, Mexico: Geological Society of America Bulletin, v. 74, no. 12, p. 1429–1438.

Dave, S. S., 1972, The geology of the igneous complex of the Barda Hills, Saurashtra, Gujarat State (India), *in* International Symposium on Deccan Trap and Other Flood Eruptions: Proceedings, Part I, Bulletin Volcanogique, v. 35, no. 3, p. 619–632.

Davino, A., 1980, The laccolith of the Tambau diabase; Determination by electrical sounding: Anais do Congresso 31, v. 5, p. 2583–2589.

Davis, W. M., 1925, Laccoliths and sills [abs.]: Washington Academy of Science Journal, v. 15, no. 18, p. 414–415.

Delaney, P. T., and Pollard, D. D., 1981, Deformation of host rocks and flow of magma during growth of minette dikes and breccia-bearing intrusions near Ship Rock, New Mexico: U.S. Geological Survey Professional Paper 1202,

61 p.

Deneufbourg, G., 1969, Geology of the Peridotite massif of southern New Caledonia; Observations: France, Bureau de Recherches Geologiques et Minieres, Bulletin, Serie 2, Section 4, no. 1, p. 27–45.

Derweiss, V., 1903, Sur les laccolites du flanc nord of la chaine de Caucase: Impromeur-librare des comptes rendus des seances de l'Academe des Sciences, Gauthier-Villars, Paris, p. 3.

de Waard, D., 1961, Tectonics of a metagabbro laccolith in the Adirondack Mountains, and its significance in determining top and bottom of a metamorphic series: Koninklijk Nederlandse Akademie Wetenschappen Proceedings, Series B, v. 64, no. 3, p. 335–342.

Didier, J., and Peyrel, J., 1980, Laccolitic structure and late Carboniferous age of Mayet de Montagne granite: Comptes Rendus Hebdomadaires des Seances de l'Academie des Sciences, Series D, Sciences Naturelles, v. 291, no. 10, p. 797–800.

Dieterich, J. H., and Decker, R. W., 1975, Finite element modeling of surface deformation associated with vulcanism: Journal of Geophysical Research, v. 80, no. 29, p. 4094–4102.

Diller, J. S., 1898, Roseburg, Oregon folio: U.S. Geological Survey Geological Atlas Folio GF-49, 4 p.

——, 1903, Port Orford, Oregon folio: U.S. Geological Survey Geological Atlas Folio GF-89, 6 p.

Dixon, H. R., 1982, Multi-stage deformation of the Preston Gabbro, eastern Connecticut, in Joestein, R., and Quarrier, S. S., eds., Guidebook for fieldtrips in Connecticut and south-central Massachusetts: Annual Meeting, New England Intercollegiage Geological Conference 74, p. 453–463.

Dockstader, D. R., 1974, Causes of regularity in sill finger geometry [abs.]: EOS American Geophysical Union Transactions, v. 55, no. 4, p. 434.

——, 1977, Fingered intrusions in a Bingham host [abs.]: EOS, Transactions, American Geophysical Union, v. 58, no. 6, p. 506.

——, 1978, The mechanics of fingered sheet intrusions [Ph.D. thesis]: Rochester, University of Rochester, 178 p.

——, 1982, Evidence for multiple intrusions in the Shonkin Sag laccolith: Geological Society of America Abstract with Programs, v. 14, no. 5, p. 258.

Douglas, J. A., 1907, On changes of physical constants which take place in certain minerals and igneous rocks, on the passage from the crystalline to the glassy state: Geological Society of London, Quarterly Journal, v. 63, p. 145–161.

Drewes, H., 1972a, Cenozoic rocks of the Santa Rita Mountains, southeast of Tucson, Arizona: U.S. Geological Survey Professional Paper 746, 66 p.

——, 1972b, Structural geology of the Santa Rita Mountains, southeast of Tucson, Arizona: U.S. Geological Survey Professional Paper 748, 35 p.

Drewes, H., and Williams, F. E., 1973, Mineral resources of the Chiricahua Wilderness area, Cochise County, Arizona: U.S. Geological Survey Bulletin 1385-A, 53 p.

Duchesne, J. C., 1978, Quantitative modeling of Sr, Ca, Rb, and K in the Bjerkreim-Sogndal layered lopolith (southwest Norway): Contributions to Mineralogy and Petrology, Beitreische Mineralogy and Petrology, v. 66, no. 2, p. 175–184.

du Toit, A., 1920, The Karroo dolerites: Transactions of the Geological Society of South Africa, v. 23, p. 1–42.

——, 1954, The geology of South Africa (third edition): Edinburgh, Oliver and Boyd, 463 p.

Eckel, E. B., 1936, Resurvey of the geology and ore-deposits of the La Plata mining district, Colorado: Proceedings of the Colorado Scientific Society, v. 13, no. 9, p. 508–547.

——, 1937, Mode of igneous intrusion in La Plata Mountains, Colorado: Transactions, American Geophysical Union, 18th Annual Meeting, part 1, p. 258–260.

Edmond, C. L., 1980, Magma immiscibility in the Shonkin Sag laccolith, Highwood Mountains, Montana [M.S. thesis]: Missoula, University of Montana, 106 p.

Ekren, E. B., and Hauser, F. N., 1956, Ute Mountains, Colorado: U.S. Geological Survey Technical Report TEI-640, p. 50–51 (also see TEI-590, p. 32–33;

TEI-620, p. 50–52; TEI-690, p. 71–75 [book 1]; TEI-700, p. 33–36; TEI-740, p. 21–29).

——, 1965, Geology and petrology of the Ute Mountains, Colorado: U.S. Geological Survey Professional Paper 481, 74 p.

El-Gaby, S., and Habib, M. E., 1978, The Bula granodiorite diapir, west of Safaga, Eastern Desert: Assist University, Faculty of Science Bulletin, v. 7, no. 2, p. 81–98.

Elston, W. E., and Snider, H. I., 1964, Differentiation and alkali metasomatism in dike swarm complex and related igneous rocks near Capitan, Lincoln County, New Mexico, in New Mexico Bureau of Mines and Mineral Resources: New Mexico Geological Society, 15th Field Conference, Guidebook, p. 140–147.

El Tahlawi, M. R., 1974, Domal structures in the Eocene plateaus surrounding the Nile Valley, Upper Egypt: Neues Jahrbuch fuer Geologie und Palaontologie, Monatshefte, no. 5, p. 257–265.

Elwood, M. W., 1976, Flow structures as a guide to intrusive form and distribution of mineralization in Eocene hypabyssal rocks of the Black Buttes, northwestern Black Hills, Wyoming: Geological Society of America Abstracts with Programs, v. 8, no. 6, p. 853–854.

Emery, W. B., 1916, The igneous geology of the Carrizo Mountains: American Journal of Science, 4th Series, v. 42, p. 349–363.

Emmart, L. A., 1985, Volatile transfer differentiation of the Gordon Butte magma, northern Crazy Mountains, Montana [M.S. thesis]: Missoula, University of Montana, 83 p.

Emmons, S. F., 1898, Tenmile district, Colorado: U.S. Geological Survey Special Folio GF-48, 6 p.

Emslie, R. F., 1970, The geology of the Michikamau intrusion, Labrador: Canada Geological Survey, Paper 68-57, 85 p.

Engel, A.E.J., and Engel, C. G., 1963, Metasomatic origin of large parts of the Adirondack phacoliths: Geological Society of America Bulletin, v. 74, no. 3, p. 349–352.

Engelder, T., and Sbar, M. L., 1984, Near-surface in situ stress; Introduction: Journal of Geophysical Research, v. 89, B11, p. 9321–9322.

England, A. W., Cooke, J. E., Hodge, S. M., and Watts, R. D., 1979, Geophysical investigations of Dufek intrusion: Antarctic Journal of the United States, v. 14, no. 5, p. 4–5.

Enmark, T., and Nisca, D. H., 1982, The Gallejaur intrusion in northern Sweden; A geophysical study, in Symposium on the geology of the Skellefte field; Summaries of talks: Geologiska Föreningen i Stockholm Förhandlinger, v. 104, no. 4, p. 381.

Erdmannsdörfer, O. H., 1924, Grundlagen der Petrographie: Stuttgart, F. Enke, 327 p.

Escher, A., 1966, The deformation and granitisation of Ketildian rocks in the Nanortalik area, south Greenland: Copenhagen, Meddelelser om Grønland, v. 172, no. 9, p 68–73.

Fairbault, E. R., Gwillim, J. C., and Barlow, A. E., 1911, Report on the geology and mineral resources of the Chibougamau region, Quebec, Quebec Province: Canada, Department of Colonization, Mines, and Fisheries, Mines Branch, 215 p.

Ferguson, H. G., Muller, S. W., and Cathcart, S. H., 1954, Geology of the Mina Quadrangle, Nevada: U.S. Geological Survey Geological Quadrangle Map GQ-45, scale 1:125,000.

Fernandez, A. 1977, Structure and emplacement of porphyritic granite at Montwert bridge, Mont Lozere, French Central Massif: Compte Rendu Sommaire des Seances de la Societe Geologique de France 3, Supplement au Bulletin 1977, tome XIX, no. 3, p. 137–140.

Fisher, J. K., 1969, Geology and structure of the Citadel Rock area, northern Black Hills, South Dakota [M.S. thesis]: Rapid City, South Dakota School of Mines and Technology, 58 p.

Foland, K. A., and Loiselle, M. C., 1981, Oliverian syenites of the Pliny region, northern New Hampshire: Geological Society of America Bulletin, v. 92, no. 4, part 1, p. 179–188.

Foose, R. M., Wise, D. U., and Garbarini, G. S., 1961, Structural geology of the Beartooth Mountains, Montana and Wyoming: Geological Society of America Bulletin, v. 72, no. 8, p. 1143–1172.

Ford, A. B., Reynolds, R. L., Huie, C., and Boyer, S. J., 1979, Geological field investigation of Dufek intrusion: Antarctic Journal of the United States, v. 14, no. 5, p. 9–11.

Friedman, M., Handin, J., Logan, J. M., Min, K. D., and Stearns, D. W., 1976, Experimental folding of rocks under confining pressure, Part III, Faulted drape folds in multilithologic layered specimens: Geological Society of America Bulletin, v. 87, no. 7, p. 1049–1066.

Fujii, N., and Uyeda, S., 1974, Thermal instabilities during flow of magma in volcanic conduits: Journal of Geophysical Research, v. 79, no. 23, p. 3367–3369.

Gandhi, S. S., and Prasad, N., 1980, Uranium and thorium variations in two monzonitic laccoliths, East Arm of Great Slave Lake, District of Mackenzie: Canada Geological Survey Paper 80-1B, p. 233–240.

—— , 1982, Comparative petrochemistry of two cogenetic monzonitic laccoliths and genesis of associated uraniferous actinolite-apatite-magnetite veins, East Arm of Great Slave Lake, District of Mackenzie, *in* Maurice, Y. T., ed., Uranium in granites: Canada Geological Survey Paper 81-23, p. 81–90.

Gaskill, D. L., and Godwin, L. H., 1966, Geologic map of the Marcellina Mountain Quadrangle, Gunnison County, Colorado: U.S. Geological Survey Geologic Quadrangle Map GQ-511, scale 1:24,000.

Gaskill, D. L., Mutschler, F. E., and Bartleson, B. L., 1981, West Elk volcanic field, Gunnison and Delta Counties, Colorado, *in* Epis, R. C., and Callender, J. F., eds., Western slope, Colorado; Western Colorado and eastern Utah: New Mexico Geological Society, 32nd Field Conference, p. 305–315.

Geikie, A., 1888, The history of volcanic action during the Tertiary period in the British Isles: Transactions of the Royal Society, Edinburgh, v. 35, p. 21–184.

Gerasimov, I. P., 1974. "Laccoliths" in Pyatigorsk and origin of mineral waters in the Caucasus: Geomortologiya, v. 3, p. 3–13.

Getz, R. C., 1965, Jointing and stratigraphy on Elkhorn Peak, Whitewood, South Dakota, and Green Mountain, Sundance, Wyoming [M.S. thesis]: Rapid City, South Dakota School of Mines and Technology, 70 p.

Gevers, T. W., 1963, Geology along the northwestern margin of the Khomas Highlands between Otjimbinque-Karibib and Okahandja, South–West Africa: Transactions of the Geological Society of South Africa, v. 66, p. 199.

Gevers, T. W., and Frommurze, H. F., 1929, The geology of northwest Damaraland, in South-West Africa: Transactions of the Geological Society of South Africa, v. 32, p. 31–55.

Gilbert, G. K., 1877, Geology of the Henry Mountains, Utah: U.S. Geographical and Geological Survey of the Rocky Mountain Region, 170 p.

—— , 1896, Laccolites in southeastern Colorado: Journal of Geology, v. 4, p. 816–825.

Gilbert, G. K., and Cross, C. W., 1896, A new laccolite locality in Colorado and its rocks [abs.]: Science, American Association for the Advancement of Science, v. 3, no. 71, p. 714.

Glangeaud, L., 1934, Etude petrographique et mineralogie du laccolite post-Burdigalien du Djebel Arroudjan (province d'Alger): Geological Society of France Bulletin, serie 5, tome fasc. 5–6, p. 367.

Godwin, L. H., 1968, Geologic map of the Chair Mountain Quadrangle, Gunnison and Pitkin Counties, Colorado: U.S. Geological Survey Quadrangle Map GQ–704, scale 1:24,000.

Godwin, L. H., and Gaskill, D. L., 1964, Post-Paleocene West Elk laccolithic cluster, west-central Colorado: U.S. Geological Survey Professional Paper 501-C, p. C66–C68.

Goodspeed, G. E., 1946, Genetic relations of magnetite deposits of the Running Wolf district in the Little Belt Mountains, Montana [abs.]: Geological Society of America Bulletin, v. 57, no. 12, part 2, p. 1252.

Gould, L. M., 1925a, Petrology of some Sierra La Sal dikes [abs.]: Pan-American Geologist, v. 44, no. 2, p. 158.

—— , 1925b, A "laccolite in the air" (La Sal Mountains, Utah): Michigan Academy of Science Papers, v. 5, p. 253–256.

—— , 1926a, Geology of the La Sal Mountains of Utah: Michigan Academy of Science Papers, v. 7, p. 55–106.

—— , 1926b, The role of orogenic stresses in laccolithic intrusions: American

Journal of Science, 5th Series, v. 12, p. 119–129.

Grant, T. C., 1979, Geology of the Spry Intrusion, Garfield County, Utah [M.S. thesis]: Kent, Kent State University, 59 p.

Grant, T. C., and Anderson, J. J., 1979, Geology of the Spry Intrusion, Garfield County, Utah: Utah Geology, v. 6, no. 2, p. 5–24.

Green A. H., 1879, The geology of the Henry Mountains: Nature, v. 21, p. 177.

Greenwood, R., and Lynch, V. M., 1959, Geology and gravimetry of the Mustang Hill laccolith, Uvalde County, Texas: Geological Society of America Bulletin, v. 70, no. 7, p. 807–825.

Gretener, P. E., 1969, On the mechanics of the intrusion of sills: Canadian Journal of Earth Science, v. 6, no. 6, p. 1415–1419.

Grout, F. F., 1918, The lopolith, an igneous form exemplified by the Duluth gabbro: American Journal of Science, 4th Series, v. 46, p. 516–522.

Grunwald, R. R., 1970, Geology and mineral deposits of the Galena mining district, Black Hills, South Dakota [Ph.D. thesis]: Rapid city, South Dakota School of Mines and Technology, 188 p.

Guerassimov, A, 1937, Les laccolithes de Piatigorsk et le versant oriental du Bechtaou: International Geological Congress, U.S.S.R., 17th session, Excursion au Caucase, p. 59–67.

Gunning, H. C., 1932, Form and mechanics of intrusion of the Nimpkish "batholith": Transactions of the Royal Society of Canada, Third Series, v. 26, Section 4, p. 289–304.

Gurbanov, A. G., and Favorskya, M. A., 1978, The problem of the neointrusions of the Caucasus, using modern data: International Geology Review, v. 20, no. 5, p. 495–506.

Hague, A., 1896, Yellowstone National Park folio, Wyoming: U.S. Geological Survey Atlas Folio GF-30, 6 p.

—— , 1898, Absaroka folio: U.S. Geological Survey Geological Atlas Folio GF-52, 6 p.

Hague, A., Iddings, J. P., Weed, W. H., Walcott, C. D., Girty, G. H., Stanton, T. W., and Knowlton, F. H., 1899, Geology of the Yellowstone National Park, Part 2: U.S. Geological Survey Monograph 32, 893 p.

Haisler, W. E., and Corry, C. E., 1983, MAGGIE User's Manual; A materially and geometrically nonlinear finite element program for static and dynamic analysis of one, two, and three dimensional structures: Nonlinear Analysis, Incorporated, c/o Department of Aerospace Engineering, Texas A&M University, College Station, Texas 77843, 186 p.

Hall, A. L., and Molengraff, G.A.F., 1925, The Vredefort mountain land in the southern Transvaal and the northern Orange Free State: Amsterdam, Utgave van de Koninklijke Academie van Wetenschappen, 183 p.

Halvorson, D. L., 1980, Geology and petrology of the Devils Tower, Missouri Buttes, and Brookes Canyon area, Crook County, Wyoming [Ph.D. thesis]: Grand Forks, University of North Dakota, 218 p.

Hamilton, W. B., 1959, Chemistry of granophyres from Wichita lopolith, Oklahoma: Geological Society of America Bulletin, v. 70, no. 8, p. 1119–1125.

Hamlyn, P. R., 1980, Equilibration history and phase chemistry of the Panton sill, Western Australia: American Journal of Science, v. 280, no. 7, p. 631–668.

Handin, J., 1966, Strength and ductility, *in* Clark, S. P., ed., Handbook of Physical Constants: Geological Society of America Memoir 97, p. 223–289.

Harker, A., 1904, Tertiary igneous rocks of Skye: United Kingdom Geological Survey Memoir, 481 p.

Harrelson, D. W., 1981, Petrography of a subsurface Triassic granite in Grenada County, Mississippi: Geological Society of America Abstracts with Programs, v. 13, no. 1, p. 9.

Harrison, J. B., 1909, Geology of the goldfields of British Guiana: London, Dulay and Company, 320 p.

Hawkes, D. D., 1966, Differentiation of the Tumatumari-Kopinang dolerite intrusion, British Guiana: Geological Society of America Bulletin, v. 77, no. 10, p. 1131–1158.

Hawkes, L., and Hawkes, H. K., 1933, The Sandfell laccolith and "dome of elevation" (Iceland): Geological Society of London, Quarterly Journal, v. 89, no. 356, part 4, p. 379–398.

Hearn, B. C., Jr., Pecora, W. T., and Swadley, W. C., 1964, Geology of the Rattlesnake Quadrangle, Bearpaw Mountains, Blaine County, Montana:

U.S. Geological Survey Bulletin 1181-B, p. B1–B66.

Heaton, R. S., 1939, Geology of Green Mountain dam site, Colorado–Big Thompson project: Colorado Society of Engineering Bulletin, Denver, v. 23, no. 11, p. 4–6 and 200.

Heidt, J. H., 1977, Geology of the Mount Theodore Roosevelt–Maitland area, Lawrence County, South Dakota [M.S. thesis]: Rapid City, South Dakota School of Mines and Technology, 141 p.

———, 1981, The Mount Theodore Roosevelt Cutting complexes, Maitland area, South Dakota: Geological Society of America Abstracts with Programs, v. 13, no. 4, p. 199.

Henry, C. D., McDowell, F. W., Price, J. G., and Smyth, R. C., 1986, Compilation of postassium-argon ages of Tertiary igneous rocks, Trans-Pecos Texas: Austin, University of Texas, Bureau of Economic Geology, Geological Circular 86-2, 34 p.

Herrin, E. T., Jr., 1957, Geology of the Solitario area, Trans-Pecos Texas [Ph.D. thesis]: Cambridge, Harvard University, 162 p.

Hess, H. H., 1960, Stillwater igneous complex, Montana: Geological Society of America Memoir 80, 230 p.

Hewitt, D. F., 1956, Geology and mineral resources of the Ivanpah Quadrangle, California and Nevada: U.S. Geological Survey Professional Paper 275, 172 p.

Hill, J. M., 1913, Notes on the northern La Sal Mountains, Grand County, Utah: U.S. Geological Survey Bulletin 530A, p. 99–118.

Hill, R., 1949, The plastic yielding of notched bars under tension: Quarterly Journal of Mechanics Applied Mathematics, v. 2, p. 40–52.

Hills, R. C., 1889a, Preliminary notes on the eruption of the Spanish Peaks region: Colorado Scientific Society, Proceedings, v. 3, no. 2, p. 24–34.

———, 1889b, Additional notes on the eruptions of the Spanish Peaks region: Colorado Scientific Society, Proceedings, v. 3, no. 2, p. 224–227.

Hirokawa, O., 1970, Investigations of titaniferous iron-ore deposits in the Wadi Hayyan–Wadi Qabqab district: Saudi Arabia, Director General of Mineral Resources, Mineral Resource Research 1969, p. 67–73.

Hirsch, T. L., and Hyndman, D. W., 1985, Evolution of Square Butte laccolith, eastern Highwood Mountains, Montana: Northwest Geology, v. 14, p. 17–31.

Hodgson, C. J., Bailes, R. J., and Verzosa, R. S., 1976, Cariboo-Bell: Canadian Institute of Mining and Metallurgy, Special Volume no. 15, p. 388–396.

Hoffman, P. F., 1968, Stratigraphy of the Lower Proterozoic (Aphebian), Great Slave Supergroup, East Arm of Great Slave Lake, District of Mackenzie: Canada, Geological Survey Paper 68-42, 93 p.

Högbom, A. G., 1909, The igneous rocks of Ragunda, Alnö, Rödö, and Nordingrå: Geologiska Föreningens i Stockholm Förhandlingar, v. 31, no. 5, p. 347–375.

Holmes, W. H., 1877, Report on the San Juan district: U.S. Geological Survey Territory (Hayden), Annual Report, v. 9, p. 237–276.

Honkala, A., 1949, A study of Tertiary intrusives and associated mineralizations in the Pillar Peak vicinity, Black Hills, South Dakota [M.S. thesis]: Lincoln, University of Nebraska, 65 p.

Horrall, K. B., 1967, Petrology of the Henderson Mountain laccolith, Park County, Montana [M.S. thesis]: DeKalb, Northern Illinois, 85 p.

Huffington, R. M., 1943, Geology of the northern Quitman Mountains, Trans-Pecos Texas: Geological Society of America Bulletin, v. 54, no. 7, p. 987–1047.

Hunt, C. B., 1938, Form of intrusion in the Henry Mountains, Utah [abs.]: Geological Society of America, Proceedings (1937), p. 88.

———, 1946, Guidebook to the geology and geography of the Henry Mountains region, Utah: Utah Geological Society Guidebook no. 1, 51 p.

———, 1958, Structural and igneous geology of the La Sal Mountains, Utah: U.S. Geological Survey Professional Paper 294-I, p. 305–364 (summary in Science, v. 119, no. 3093, p. 477–478, 1954; reprinted in Utah Geological Association Publication 8, p. 25–106, 1980).

———, 1980, G. K. Gilbert, on laccoliths and intrusive structures, in Yochelson, E. L., ed., The scientific ideas of G. K. Gilbert: Geological Society of America Special Paper 183, p. 25–34 (also Utah Geological Association Publication 8, p. 15–24).

———, 1983, Development of the La Sal and other laccolithic mountains on the Colorado Plateau, in Averett, W. R., ed., Northern Paradox Basin—Uncompaghre uplift: Grand Junction Geological Society, Guidebook, p. 29–32.

Hunt, C. B., and Mabey, D. R., 1966, Stratigraphy and structure, Death Valley, California: U.S. Geological Survey Professional Paper 494A, 162 p.

Hunt, C. B., Averitt, P., and Miller, R. L., 1953, Geology and geography of the Henry Mountains region, Utah: U.S. Geological Survey Professional Paper 228, 234 p.

Huppert, H. E., and Sparks, R.S.J., 1984, Double-diffusive convection due to crystallization in magmas: Annual Review of Earth and Planetary Science, v. 12, p. 11–37.

Hurlburt, C. S., Jr., 1936, Differentiation in the Shonkin Sag laccolith, Montana [abs.]: American Mineralogist, v. 21, no. 3, p. 198.

Hurlburt, C. S., Jr., and Griggs, D. T., 1939, Igneous rocks of the Highwood Mountains, Montana, Part 1, The laccoliths: Geological Society of America Bulletin, v. 50, no. 7, p. 1045–1112.

Hyndman, D. W., and Alt, D., 1982, Laccoliths with exposed feeder dikes in the Montana alkalic province: Geological Society of America Abstracts with Programs, v. 14, p. 315.

———, 1983, Emplacement of alkalic intrusives, central Montana: Geological Society of America Abstracts with Programs, v. 15, no. 5, p. 420.

———, 1987, Radial dikes, laccoliths, and gelatin models: Journal of Geology, v. 95, p. 763–774.

Iddings, J. P., 1898, Bysmaliths: Journal of Geology, v. 6, p. 704–710.

Iddings, J. P., and Weed, W. H., 1894, Livingston, Montana: U.S. Geological Survey Geological Atlas Folio GF-1, 5 p.

———, 1899, Descriptive geology of the Gallatin Mountains, in Geology of the Yellowstone National Park: U.S. Geological Survey Monograph 32, part 2, p. 1–59.

Indest, S., and Carman, M., 1979, Crystallization history of the Wildhorse Mountain quartz syenite intrusion and its relation to some other Big Bend intrusions, in Walton, A. W., and Henry, C. D., eds., Cenozoic geology of the Trans-Pecos volcanic field of Texas: Austin, University of Texas, Bureau of Economic Geology, Guidebook 19, p. 72–82.

Ingerson, F. E., 1935, Layered peridotitic laccoliths of the Trout River area, Newfoundland: American Journal of Science, 5th Series, v. 29, p. 422–440.

———, 1937, Layered peridotite laccoliths in the Trout River area, Newfoundland; A reply: American Journal of Science, 5th Series, v. 33, p. 389–392.

Irving, J. D., 1899, A contribution to the geology of the northern Black Hills: New York Academy of Science Annals, v. 12, p. 187–340.

Jackson, M. D., 1987, Deformation of host rocks during growth of igneous domes, southern Henry Mountains, Utah [Ph.D. thesis]: Baltimore, Johns Hopkins University, 155 p.

Jackson, M. D., and Pollard, D. D., 1988, The laccolith-stock controversy: New results from the southern Henry Mountains, Utah: Geological Society of America Bulletin, v. 100, no. 1, p. 117–139.

Jaeger, J. C., 1957, The temperature in the neighborhood of a cooling intrusive sheet: American Journal of Science, v. 255, no. 4, p. 306–318.

———, 1959, Temperature outside a cooling intrusive sheet: American Journal of Science, v. 257, no. 1, p. 44–54.

———, 1962, Punching tests on disks of rock under hydrostatic pressure: Journal of Geophysical Research, v. 67, no. 1, p. 369–373.

———, 1968, Cooling and solidification of igneous rocks, in Hess, H. H., ed., Basalts, The Poldervaart treatise on rocks of basaltic composition: New York, Wiley Interscience, v. 2, p. 503–536.

Jaeger, J. C., and Cook, N.G.W., 1976, Fundamentals of rock mechanics (second edition): London, Chapman and Hall, Limited, 585 p.

Jaggar, T. A., 1901, The laccoliths of the Black Hills, with a chapter on Intrusion and erosion, by Ernest Howe: U.S. Geological Survey, 21st Annual Report, part 3, p. 163–290.

——, 1904, Economic resources of the northern Black Hills: U.S. Geological Survey Professional Paper 26, Part I, p. 7–41.

James, R. S., 1984, Geology and geochemistry of the East Bull Lake layered intrusion, Ontario: Geological Association of Canada Program with Abstracts, v. 9, p. 76.

James, R. S., and Born, P., 1985, Geology and geochemistry of the East Bull Lake intrusion, District of Algoma, Ontario: Canadian Journal of Earth Sciences, v. 22, no. 7, p. 968–979.

Jevons, H. S., Jensen, H. I., Taylor, T. G., and Süssmilch, C. A., 1911, The geology and petrography of the Prospect intrusion: Proceedings of the Royal Society of New South Wales, v. 45, p. 445–553.

——, 1912, The differentiation phenomena of the Prospect intrusion: Proceedings of the Royal Society of New South Wales, v. 46, p. 111–138.

Johnson, A. M., 1970, Physical processes in geology: San Francisco, Freeman, Cooper and Company, 577 p.

Johnson, A. M., and Ellen, S. D., 1974, A theory of concentric, kink, and sinusoidal folding and of monoclinal flexuring of compressible, elastic multilayers, Part I, Introduction: Tectonophysics, v. 21, no. 4, p. 301–339.

Johnson, A. M., and Pollard, D. D., 1973, Mechanics of growth of some laccolithic intrusions in the Henry Mountains, Utah, Part I, Field observations, Gilbert's model, physical properties, and flow of the magma: Tectonophysics, v. 18, no. 3–4, p. 261–309.

Johnson, B. K., 1957, Geology of a part of the Manly Peak Quadrangle, southern Panamint Range, California: Berkeley, California University Publications in Geological Sciences, v. 30, no. 5, p. 353–423.

Johnson, R. B., 1961, Coal resources of the Trinidad coal field in Huerfano and Las Animas Counties, Colorado: U.S. Geological Survey Bulletin, 1112-E, p. 129–180.

——, 1968, Geology of the igneous rocks of the Spanish Peaks region, Colorado: U.S. Geological Survey Professional Paper 594-G, 47 p.

Jones, W. R., Hernon, R. M., and Moore, S. L., 1967, General geology of Santa Rita Quadrangle, Grant County, New Mexico: U.S. Geological Survey Professional Paper 555, 144 p.

Jones, W. R., Hernon, R. M., and Pratt, W. P., 1961, Geologic events culminating in primary metalization in the Central mining district, Grant County, New Mexico: U.S. Geological Survey Professional Paper 424-C, p. C11–C16.

Jordan, T. H., Creager, K. C., and Fischer, K. M., 1986, Subduction flow and mantle dynamics [abs.]: EOS, Transactions, American Geophysical Union, v. 67, no. 16, p. 379.

Jumnogthai, M., 1979, Geology of the Sugarloaf Mountain area, Lead, South Dakota [M.S. thesis]: Rapid City, South Dakota School of Mines and Technology, 101 p.

Junger, A., and Sylvester, A. G., 1979, Origin of Emery Seaknoll, southern California borderland [abs.]: EOS, Transactions, American Geophysical Union, v. 60, no. 46, p. 951.

Kane, M. F., and Bromery, R. W., 1968, Gravity anomalies in Maine, *in* Zen, E., White, W. S., Hadley, J. B., and Thompson, J. B., eds., Studies of Appalachian geology, northern and maritime: New York, Interscience Publications, p. 415–423.

Kantsel, A. V., Korotayev, M. A., Laverov, N. P., and Smorchkov, I. Ye., 1972, The Babaytag subvolcanic massif in Central Asia; Its geological structure and history of formation, *in* Smorchkov, I. Ye., ed., Geologiya liparitovoy formatsii rayonov Sredney Azii i Kazakhstana: Izdatelstvo Nauka, Moscow, SUN., p. 108–122.

Karlo, J. F., and Jorgenson, D. B., 1979, Fault control of volcanic features southeast of Blackfoot, Snake River plain, Idaho: Geological Society of America Abstracts with Programs, v. 11, no. 6, p. 276.

Kebuladze, V. V., and Tatishvili, O. V., 1969, Magnetotelluric survey of the semi-metallic deposits of Adzharistan: Akademiya Nauk Gruzinskoy S.S.R., Soobshcheniya, v. 54, no. 2, p. 325–328.

Kelley, V. C., 1945, Geology, ore deposits, and mines of the Mineral Point, Poughkeepsie, and upper Uncompaghre districts, Ouray, San Juan, Hinsdale Counties, Colorado: Colorado Scientific Society Proceedings, v. 14, no. 7, p. 287–446.

——, 1946, Stratigraphy and structure of the Gallinas Mountains, New Mexico [abs.]: Geological Society of America Bulletin, v. 57, no. 12, part 2, p. 1254.

Kendrick, G. C., 1980a, Magma immiscibility in the Square Butte laccolith of central Montana [M.S. thesis]: Missoula, University of Montana, 90 p.

——, 1980b, Field relationships in the Square Butte laccolith of central Montana: Northwest Geology, v. 9, p. 26–34.

Kendrick, G. C., and Edmond, C. L., 1981, Magma immiscibility in the Shonkin Sag and Square Butte laccoliths: Geology, v. 9, no. 12, p. 615–619 (also see discussion and reply in v. 10, no. 8, p. 444–446).

Kennan, P. S., 1979, Plutonic rocks in the Irish Caledonides, *in* Harris, A. L., Holland, C. H., and Leake, B. E., eds., The Caledonides of the British Isles; Reviewed: Geological Society of London Special Publication 8, p. 705–711.

Kerr, J. H., Pecora, W. T., Stewart, D. B., and Dixon, H. R., 1957, Preliminary geologic map of the Shambo Quadrangle, Bearpaw Mountains, Montana: U.S. Geological Survey Miscellaneous Geologic Investigations Map I-236, scale 1:31,680.

Kettner, R., 1914, Über die Lakkolithenartigen intrusionen der porphyre zwischen Mnisek und der Moldau: International de L'Academy des Science de Boheme Bulletin, p. 73–97.

Keyes, C. R., 1918, Mechanics of laccolithic intrusion [abs.]: Geological Society of America Bulletin, v. 29, p. 75 (also see Revue de Geologie, Annal 4, no. 1, p. 39–40, 1923).

——, 1922, New Mexican laccolithic structures: Pan-American Geologist, v. 37, no. 2, p. 109–120.

Khasanov, R. Kh., 1968, The form of the Lyangar intrusive, southeastern Pamirs: Akademiya Nauk Tadzhikshoy S.S.R., Doklady, v. 11, no. 4, p. 48–51.

Kilinc, I. A., 1979, Melting relations in the quartz diorite–H_2O and quartz diorite H_2O–CO_2 systems: Neues Jahrbuch fur Mineralogie, Monatshefte, v. 2, p. 67–72.

Kingma, J. T., 1974, The geological structure of New Zealand: New York, John Wiley and Sons, 407 p.

Kirchner, J. G., 1971, The petrography and petrology of the phonolite porphyry intrusions of the northern Black Hills, South Dakota [Ph.D. thesis]: Ames, University of Iowa, 199 p.

Kistler, R. W., and Ford, A. B., 1979, Potassium-argon ages of Dufek intrusion and other Mesozoic mafic bodies in the Pensacola Mountains: Antarctic Journal of the United States, v. 14, no. 5, p. 8–9.

Kleinkopf, M. D., 1970, Magnetic and gravity studies in the Black Hills, South Dakota: Geological Society of America Abstracts with Programs, v. 2, no. 5, p. 338–339.

——, 1980, Aeromagnetic studies of the Square Butte Wilderness Area, Choteau County, Montana: U.S. Geological Survey Professional Paper 1175, p. 2.

Kleinkopf, M. D., and Redden, J. A., 1975, Bouguer gravity, aeromagnetic, and generalized geologic map of the Black Hills of South Dakota and Wyoming: U.S. Geological Survey Geophysical Investigations Map GP-903, scale 1:250,000, 11 p.

Kleinkopf, M. D., Witkind, I. J., and Keefer, W. R., 1972, Aeromagnetic, Bouguer gravity, and generalized geologic maps of the central part of the Little Belt Mountains, Montana: U.S. Geological Survey Geophysical Investigations Map GP-837, scale 1:62,500, 4 p.

Klepper, M. R., Weeks, R. A., and Ruppel, E. T., 1957, Geology of the southern Elkhorn Mountains, Jefferson and Broadwater Counties, Montana: U.S. Geological Survey Professional Paper 292, 82 p.

Klepper, M. R., Ruppel, E. T., Freeman, V. L., and Weeks, R. A., 1971, Geology and mineral deposits, east flank of the Elkhorn Mountins, Broadwater County, Montana: U.S. Geological Survey Professional Paper 665, 66 p.

Knechtel, M. M., 1959, Stratigraphy of the Little Rocky Mountains and encircling foothills, Montana: U.S. Geological Survey Bulletin 1072-N, p. 723–752.

Knight, G. L., and Landes, K. K., 1932, Kansas laccoliths: Journal of Geology, v. 40, no. 1, p. 1–15.

Koch, F. G., Johnson, A. M., and Pollard, D. D., 1981, Monoclinal bending of strata over laccolithic intrusions: Tectonophysics, v. 74, no. 3–4, p. T21–T31.

Kolessar, J., 1970, Geology and copper deposits of the Tyrone district, *in* Guiebook of the Tyrone–Big Hatchet Mountains–Florida Mountains region: New Mexico Geological Society, 21st Field Conference, p. 127–132.

——, 1982, The Tyrone copper deposit, Grant County, New Mexico, *in* Titley, S. R., ed., Advances in geology of porphyry copper deposits; Southwestern North America: Tucson, University of Arizona Press, p. 327–333.

Kovschak, A. A., Jr., 1974, Igneous and structural geology of the Grapevine Hills, Big Bend National Park, Brewster County, Texas [M.S. thesis]: Arlington, University of Texas, 107 p.

——, 1976, Igneous and structural geology of the Grapevine Hills, Big Bend National Park, Brewster County, Texas: Geological Society of America Abstracts with Programs, v. 8, no. 1, p. 27.

Krause, H., and Pape, H., 1977, The geology and petrography of the Storgangen ilmenite ore body and its wallrock units, southern Norway: Norsk Geologisk Tidsskrift, v. 57, no. 3, p. 263–284.

Kravchenko, S. M., and Beskin, S. M., 1979, Cryptic layering of the Akzhaylyautas granite lopolith: Doklady of the U.S.S.R., Earth Science Sections, v. 244, no. 1-6, p. 95–98.

Kreiger, E. W., Jr., 1976, Geology and petrology of the Two Buttes intrusion [Ph.D. thesis]: University Park, Pennsylvania State University, 116 p.

Kreiger, E. W., Jr., and Thorton, C. P., 1976, Geology and petrology of the Two Buttes intrusion: Geological Society of America Abstracts with Programs, v. 8, no. 5, p. 596–597.

Kuhn, P. W., 1982, Magma immiscibility in the Box Elder laccolith, westmost Bearpaw Mountains, north-central Montana: Geological Society of America Abstracts with Programs, v. 14, no. 6, p. 318.

——, 1983, Magma immiscibility in the Box Elder laccolith of north-central Montana [M.S. thesis]: Missoula, University of Montana, 86 p.

Kulcsar, L., and Barta, I., 1971, Petrographic study of a laccolith near Erdobenye, Hungary: Acta Geographica Debrecina 1969–70, v. 16–17, ser. 8–9, p. 39–72.

Lachmann, R., 1909, Der eruptions mechanisms bei den Euganeetrachyten: Zeitschrift der Deutschen Geologischen Gesellschaft, Monatsberichte Band 61, p. 331–340.

Lapworth, C., and Watts, W. W., 1894, The geology of south Shropshire: Geologists' Association of London, Proceedings, v. 13, p. 297–355.

Larsen, E. S., Jr., 1941, Igneous rocks of the Highwood Mountains, Montana, Part 2, The extrusive rocks: Geological Society of America Bulletin, v. 52, no. 11, p. 1733–1751 (also see Geological Society of America Bulletin, v. 52, no. 12, p. 1829–1868).

Larsen, E. S., Jr., and Cross, W., 1956, Geology and petrology of the San Juan region, southwestern Colorado: U.S. Geological Survey Professional Paper 258, 303 p.

Larsen, E. S., Jr., Hurlburt, C. S., Jr., Burgess, C. H., Griggs, D. T., and Buie, B. F., 1935, The igneous rocks of the Highwood Mountains of central Montana: Transactions, American Geophysical Union, National Research Council, 16th Annual Meeting, p. 288–292.

Larsen, L. H., and Simms, F. E., 1972, Igneous geology of the Crazy Mountains, Montana; A report of work in progress: Montana Geological Society, Annual Field Conference Guidebook no. 21, p. 135–139.

Laurén, L., 1970, An interpretation of the negative gravity anomalies associated with the Rapakivi granites and the Jotnian Sandstone in southern Finland: Geoligiska Föreningins i Stockholm Förhandlingar, v. 92, p. 21–34.

Lawson, A. C., 1893, The laccolitic sills of the northwest coast of Lake Superior: Minnesota Geological Survey, Bulletin 8, part 2, p. 24–48.

——, 1914, Is the Boulder "batholith" a laccolith? A problem in ore genesis: Bulletin of the Department of Geology, University of California Publications, v. 8, no. 1, p. 1–15.

Lazebnik, K. A., and Lazebnik, Yu. D., 1979, Potassium-richterite in metasomatic rocks of Murunsk Massif, *in* Rundkvist, D. V., ed., Mineraly i parageneisy mieralov gornykh porod i rud: Lenigrad, Izdatelstvo Nauka, U.S.S.R., p. 119–122.

Leaman, D. E., and Naqui, I. H., 1967, Geology and geophysics of the Cygnet district: Tasmania Geological Survey Bulletin no. 49, 110 p.

Leeman, W. P., and Gettings, M. E., 1977, Holocene rhyolite in southeast Idaho

and geothermal potential [abs.]: EOS, Transactions, American Geophysical Union, v. 58, no. 12, p. 1249.

Leith, C. K., and Harder, E. C., 1908, Iron ores of the Iron Springs district, southern Utah: U.S. Geological Survey Bulletin 338, 102 p.

Leo, G. W., 1979, The Oliverian domes, re-evaluated: U.S. Geological Survey Professional Paper 1150, p. 60.

——, 1980a, Petrology and geochemistry of the Oliverian domes of New England: Geological Survey Professional Paper 1175, p. 200.

——, 1980b, Petrology and geochemistry of Oliverian core gneisses; A progress report: Geological Society of America Abstracts with Programs, v. 12, no. 2, p. 69.

Leppert, D. E., 1985, Differentiation of a shoshonitic magma at Snake Butte, Blaine County, Montana [M.S. thesis]: Missoula, University of Montana, 121 p.

Leveratto, M. A., 1968, Geology of the zone west of Ullun-Zonda, eastern border of the San Juan precordillera, subvolcanic activity and structure: Associacion Geologica Argentina, Revista, v. 23, no. 2, p. 129–157.

Levskiy, L. K., and Rubley, A. G., 1978, Ar 40/Ar 39 dating of young rocks: International Geology Review, v. 20, no. 4, p. 435–440.

Lewis, C. D., Jr., 1976, A paleomagnetic investigation of three intrusions in Big Bend National Park Brewster County, Texas [M.S. thesis]: Norman, University of Oklahoma, 105 p.

Lind, G., 1967, Gravity measurements over the Bohus granite in Sweden: Geologiska Föreningen geni Stockholm, Förhandlingar, v. 88, p. 542–548.

Lindsey, D. A., 1982, Geologic map and discussion of selected mineral resources of the North and South Moccasin Mountains, Fergus County, Montana: U.S. Geological Survey Miscellaneous Investigations I-1362, 8 p.

Lindsey, D. A., and Naeser, C. W., 1985, Relation between igneous intrusion and gold mineralization in the North Moccasin Mountains, Fergus County, Montana: U.S. Geological Survey Professional Paper 1301-B, p. 35–41.

Liptak, A. R., 1984, Geology and differentiation of Round Butte laccolith, central Montana [M.S. thesis]: Missoula, University of Montana, 49 p.

Lisenbee, A. L., 1981, Tertiary igneous province; Northern Black Hills, South Dakota and Wyoming; An overview: Geological Society of America Abstracts with Programs, v. 13, no. 4, p. 202.

Lonsdale, J. T., 1940, Igneous rocks of the Terlingua-Solitario region, Texas: Geological Society of America Bulletin, v. 51, no. 10, p. 1539–1636.

Lonsdale, P., 1983, Laccoliths(?) and small volcanoes on the flank of the East Pacific Rise: Geology, v. 11, no. 12, p. 706–709.

Loughlin, G. F., 1912, The gabbros and associated rocks at Preston, Connecticut: U.S. Geological Survey Bulletin 492, 158 p.

Love, J. D., 1939, Geology along the southern margin of the Absaroka Range, Wyoming: Geological Society of America Special Paper 20, 134 p.

Lovejoy, E.M.P., 1976, Geology of the Cerro de Cristo Rey uplift, Chihuahua and New Mexico: New Mexico Bureau of Mines and Mineral Resources Memoir 31, 84 p.

Luedke, R. G., and Burbank, W. S., 1962, Geology of the Ouray Quadrangle, Colorado: U.S. Geological Survey Quadrangle Map GQ-152, scale 1:24,000.

Lyons, J. B., 1944, Igneous rocks of the northern Big Belt Range, Montana: Geological Society of America Bulletin, v. 55, no. 4, p. 445–472.

MacCarthy, G. R., 1925, Some facts and theories concerning laccoliths: Journal of Geology, v. 33, p. 1–18.

MacDonald, G. A., 1972, Volcanoes: Englewood Cliffs, New Jersey, Prentice-Hall, 510 p.

Mackin, J. H., 1947a, Joint patterns in the Three Peaks laccolith, Iron Springs district, Utah [abs.]: Geological Society of America Bulletin, v. 58, no. 12, part 2, p. 1255.

——, 1947b, Some structural features of the intrusions in the Iron Springs district: Utah Geological Society Guidebook no. 2, 67 p. (reprinted in 1974 as Utah Geological and Mineral Survey Publication no. 98, 62 p.).

——, 1960, Structural significance of the Tertiary volcanic rocks of southwestern Utah: American Journal of Science, v. 258, p. 81–131.

——, 1968, Iron ore deposits of the Iron Springs district, southwestern Utah, *in*

Ore deposits of the United States, 1933–1967 (Graton-Sales Volume), v. 2: New York, American Institute of Mining, Metallurgy, and Petroleum Engineers, p. 992–1019.

Maguire, T. J., 1980, Gravity study of area adjacent to the Fall Zone of northwestern Delawawre and northeastern Maryland [M.S. thesis]: Newark, University of Delaware, 111 p.

Mapel, W. J., Robinson, C. S., and Theobald, P. K., 1959, Geologic and structure contour map of the northern and western flanks of the Black Hills, Wyoming, Montana, and South Dakota: U.S. Geological Survey Oil and Gas Investigations Map OM-191, scale 1:96,000.

Marsh, B. D., 1982, On the mechanics of igneous diapirism, stoping, and zone melting: American Journal of Science, v. 282, no. 6, p. 808–855.

Mason, R. , 1971, The chemistry and structure of the Sulitjelma gabbro [with discussion], *in* The Caledonian geology of northern Norway: Norges Geologiske Undersokelse, no. 269, p. 108–142.

Matthews, C. B., III, 1979, Geology of the central Vanocker laccolith area, Meade County, South Dakota [M.S. thesis]: Rapid City, South Dakota School of Mines and Technology, 111 p.

——— , 1981, The Vanocker laccolith, Meade County, South Dakota: Geological Society of America Abstracts with Programs, v. 13, no. 4, p. 219.

Maurer, W. C., 1965, Shear failure of rock under compression: Society of Petroleum Engineers Journal, v. 5, no. 2, p. 167–176.

Maxwell, R. A., Lonsdale, J. T., Hazzard, R. T., and Wilson, J. A., 1967, Geology of Big Bend National Park, Brewster County, Texas: Austin, University of Texas, Bureau of Economic Geology, Publication 6711, 320 p.

Mazzanti, R., Squarci, P., and Taffi, L., 1963, Geologia della zona di Montecatini val di Cecina in provincia di Pisa: Italian Geological Society Bulletin, v. 82, no. 2, p. 1–67.

McAnulty, W. N., 1980, Geology and mineralization of the Sierra Blanca peaks, Hudspeth County, Texas, *in* Dickerson, P. W., Hoffer, J. M., and Callender, J. F., eds., Trans-Pecos region: New Mexico Geological Society, 31st Field Conference, p. 263–266.

McBirney, A. R., 1984, Igneous petrology: San Francisco, Freeman Cooper and Company, 504 p.

McBirney, A. R., and Murase, T., 1984, Rheological properties of magmas: Annual Review of Earth and Planetary Sciences, v. 12, p. 337–357.

McBride, C. K., 1979, Small laccoliths and feeder dikes of the northern Adel Mountain volcanics: Northwest Geology, v. 8, p. 1–9.

McDowell, D., 1974, Emplacement of the Little Chief stock, Panamint Range, California: Geological Society of America Bulletin, v. 85, p. 1535–1546.

McKnight, J. F., 1963a, Igneous rocks of the Sombrero Tillo, northern Sierra de Picachos, Nueva Leon, Mexico [M.A. thesis]: Austin, University of Texas, 90 p.

——— , 1963b, Igneous rocks of the Northern Sierra de Picachos, Nuevo Leon, Mexico [abs.]: Texas Journal of Science, v. 15, no. 4, p. 410.

——— , 1970, Geology of the Bofecillos Mountains area, Trans-Pecos Texas: Austin, University of Texas, Bureau of Economic Geology, Geological Quadrangle Map 37, scale 1:48,000, 36 p.

McMillan, D. K., 1979, Crystallization and metasomatism of the Cuchillo Mountain laccolith, Sierra County, New Mexico [Ph.D. thesis]: Stanford, Stanford University, 298 p.

McMillan, D. K., and Jahns, R. H., 1979, Metasomatism in the Cuchillo Mountain laccolith, Sierra County, New Mexico: Geological Society of America Abstracts with Programs, v. 11, no. 6, p. 280.

Meier, L. F., 1979, Geology of the Crow Peak area: Geological Society of America Abstracts with Programs, v. 11, no. 6, p. 208.

Menot, R., and Piboule, M., 1977, The basic and ultrabasic premetamorphic massifs in the Chalus region, Limousin (Chalus sheet XIX-32, scale 1:50,000): France, Bureau de Recherches Geologiques et Minieres, Bulletin, Series 2, sect. 1, no. 4, p. 307–332.

Milanovsky, E. E., and Koronovsky, N. V., 1966, Ignimbrite-tufflava *in* Cook, E., ed., Tufflavas and ignimbrites: New York, American Elsevier Publishing Company, Incorporated, p. 72–87.

Miller, R. N., 1959, Geology of the South Moccasin Mountains, Fergus County,

Montana: Montana Bureau of Mines and Geology Memoir 37, 44 p.

Miller, W.J., 1929, Significance of newly found Adirondack anorthosite: American Journal of Science, 5th Series, v. 18, p. 383–400.

Min, K. D., 1974, Analytical and petrofabric studies of experimental faulted drape folds in layered rock specimens [Ph.D. thesis]: College Station, Texas A&M University, 89 p.

Minakami, T., Ishikawa, T., and Yagi, K., 1951, The 1944 eruption of Volcano Usu in Hokkaido, Japan: Volcanological Bulletin, series 2, v. 11, p. 5–157.

Mintz, Y., 1942, Slate Mountain volcano—laccolith [Arizona]: Plateau, v. 14, no. 3, p. 42–47.

Molengraff, G.A.F., 1901, Geologie de la Republique Sud-Africaine du Transvaal: Paris, Societe Geologique de France Bulletin, ser. 4, tome 1, p. 13–92.

——— , 1904, Geology of the Transvaal: Edinburgh and Johannesburg, T. and A. Constable, 49 p. (translated from the French edition).

Morse, S. A., 1969, The Kiglapait layered intrusion, Labrador: Geological Society of America Memoir 112, 204 p.

Morton, L. B., 1983, Geology of the Mount Ellen Quadrangle, Henry Mountains, Utah [M.S. thesis]: Provo, Brigham Young University (published in Brigham Young University Geology Studies, v. 31, Part 1, p. 67–96, 1984).

Mosconi, L. S., 1984, Tectonic history of the Black Mesa–Lowes Valley area, Brewster County, Texas [M.A. thesis]: Nacadoches, Stephen F. Austin State University, 72 p.

Mudge, M. R., 1968, Depth control of some concordant intrusions: Geological Society of America Bulletin, v. 79, no. 3, p. 315–332.

Mukhurjee, N. S., 1968, Geology and mineral deposits of the Galena–Gilt Edge area, northern Black Hills, South Dakota [Ph.D. thesis]: Golden, Colorado School of Mines, 153 p.

Muller, O. H., 1986, Changing stresses during emplacement of the radial dike swarm at Spanish Peaks, Colorado: Geology, v. 14, no. 2, p. 157–159.

Murase, T., and McBirney, A. R., 1973, Properties of some common igneous rocks and their melts at high temperatures: Geological Society of America Bulletin, v. 84, p. 3563–3592.

Murata, K. J., and Richter, D. H., 1961, Magmatic differentiation in the Uwekahuna laccolith, Kilauea Caldera, Hawaii: Journal of Petrology, v. 2, no. 3, p. 424–437.

Mutschler, F. E., Ernst, D. R., Gaskill, D. L., and Billings, P., 1981, Igneous rocks of the Elk Mountains and vicinity, Colorado; Chemistry and related ore deposits, *in* Epis, R. C., and Callender, J. F., eds., western slope, Colorado; western Colorado and eastern Utah: New Mexico Geological Society, 32nd Field Conference, p. 317–324.

Nagasaki, H., 1966, A layered ultrabasic complex at Horoman, Hokkaido, Japan: Tokyo University, Faculty of Science Journal, Section 2, v. 16, part 2, p. 313–346.

Nash, W. P., 1972, Apatite chemistry and phosphorous fugacity in a differentiated igneous intrusion: American Mineralogy, v. 57, no. 5–6, p. 877–886 (also see correction in v. 58, no. 3–4, p. 345, 1973).

Nash, W. P., and Kendrick, G. C., 1982, Magma immiscibility in the Shonkin Sag and Square Buttes laccoliths; Discussion and reply: Geology, v. 10, no. 8, p. 444–446 (also see original article by Kendrick, G. C., and Edmond, C. L., v. 9, no. 12, p. 615–619, 1981).

Nash, W. P., and Wilkinson, J. F., 1969, Crystallization of mafic minerals in the Shonkin Sag laccolith, Montana: Geological Society of America Abstracts with Programs, v. 2, p. 157–158.

——— , 1970, Shonkin Sag laccolith, Montana, Part I, Mafic minerals and estimates of temperature, pressure, oxygen fugacity, and silica activity: Contributions to Mineralogy and Petrology, v. 25, no. 4, p. 241–269.

——— , 1971, Shonkin Sag laccolith, Montana, Part II, Bulk rock geochemistry: Contributions to Mineralogy and Petrology, v. 33, no. 2, p. 162–170.

Naylor, R. S., 1968, Origin and regional relationships of the core rocks of the Oliverian domes, *in* Zen, E-an, White, W. S., Hadley, J. B., and Thompson, J. B., eds., Studies of Appalachian geology, Northern and maritime: New York, Interscience Publications, p. 231–240.

Newell, R. A., 1982, Geologic setting and geology of lead - zinc - silver deposits of the Tombstone district, Cochise County, Arizona: Geological Society of

America Abstracts with Programs, v. 14, no. 4, p. 220.

Nicoll, L. D., and Nicholls, J. , 1974, Petrogenesis of the Highwood Mountain area, Montana: Geological Society of America Abstracts with Programs, v. 6, no. 7, p. 890–891.

Nodop, I., 1971, Deep seismic refraction studies in the profile of Versmold-Luebbecke-Nienburg: Fortschritte in der Geologie von Rhineland und Westfalen, v. 18, p. 411–421.

Ochoterena, F., 1981, Evolution of morphostructural units in the Diguiyu region: Boletin del Institutio de Geografia, v. 10, p. 285–317.

Oinouye, Y., 1917, A few interesting phenomena on the eruption of Usu: Journal of Geology, v. 25, p. 258–288.

Omori, F., 1911a, The Usu-san eruption and earthquake and elevation phenomena: Tokyo, Japan Imperial Earthquake Investigations Committee, Bulletin, v. 5, no. 1, p. 1–38.

—— , 1911b, The Usu-san eruption and the earthquake and elevation phenomena, Part II, Comparison of the bench mark heights in the base district before and after the eruption: Tokyo, Japan Imperial Earthquake Investigations Committee, Bulletin, v. 5, no. 1, p. 101–107.

Osborne, F. F., and Roberts, E. J., 1931, Differentiation in the Shonkin Sag laccolith, Montana: American Journal of Science, 5th Series, v. 22, p. 331–353.

Osman, C. W., 1924, The geology of the northern border of Dartmoor between Whiddon Down and Butterdon Down: Quarterly Journal of the Geological Society of London, v. 80, p. 315–337.

—— , 1928, The granites of the Scilly Isles and their relation to the Dartmoor granites: Quarterly Journal of the Geological Society of London, v. 84, p. 258–292.

Page, N. J., 1977, Stillwater complex, Montana: rock succession, metamorphism, and structure of the complex and adjacent rocks: U.S. Geological Survey Professional Paper 999, 79 p.

Paige, S. , 1913, The bearing of progressive increase of viscosity during intrusion on the form of laccoliths: Journal of Geology, v. 21, p. 541–549.

—— , 1916, Description of the Silver City Quadrangle, New Mexico: U.S. Geological Survey Geological Atlas Folio GF-199, 19 p.

Palmer, H. S., 1925, Structure of the South Moccasin laccolith, Montana: American Journal of Science, 5th Series, v. 10, p. 119–133.

Parkkinen, J., 1975, Deformation analysis of a Precambrian mafic intrusive: Geological Survey of Finland Bulletin no. 278, p. 1–61.

Parsons, I., 1965, The sub-surface shape of part of the Loch Ailsh intrusion, Assynt, as deduced from magnetic anomalies across the contact, with a note on traverses across the Loch Borralan complex: Geology Magazine, v. 102, no. 1, p. 46–58.

Parsons, W. H., 1942, Origin and structure of the Livingston igneous rocks, Montana: Geological Society of America Bulletin, v. 53, no. 8, p. 1175–1185.

Parsons, W. H., and Stow, M. H., 1942, Origin and structural relationships of the igneous member of the Livingston Formation, Montana: Transactions, American Geophysical Union, 23rd Annual Meeting, Part 2, p. 344–345.

Parsons, W. H., Miller, C., St. Aubin, T., and Pozy, W., 1972, Petrology, geochemistry, and paleomagnetism of Eocene volcanic rocks of Sunlight basin, Absaroka Mountains, Wyoming: Geological Society of America Abstracts with Programs, v. 4, no. 6, p. 401.

Peale, A. C., 1896, Three Forks, Montana: U.S. Geological Survey Geological Atlas Folio GF-24, 7 p.

Pecora, W. T., 1941, Structure and petrology of the Boxelder laccolith, Bearpaw Mountains, Montana: Geological Society of America Bulletin, v. 52, no. 6, p. 817–853.

Pecora, W. T., Witkind, I. J., and Stewart, D. B., 1957a, Preliminary general geologic map of the Laredo Quadrangle, Bearpaw Mountains, Montana: U.S. Geological Survey Miscellaneous Geological Investigations Map I-234, scale 1:31,680.

Pecora, W. T., Kerr, J. H., Brace, W. F., Stewart, D. B., Engstrom, D. B., and Dixon, H. R., 1957b, Preliminary geologic map of the Warrick Quadrangle, Bearpaw Mountains, Montana: U.S. Geological Survey Miscellaneous Geo-logical Investigations Map I-237, scale 1:31,680.

Pecorini, G., 1966, Sull'eta "oligocenica" del vulcanismo al bordo orientale della fossa tectonica del Campidano (Sardegna): Accademia Nazationale dei Lincei, Atti, Classe di Scienze Fisiche, Matematiche e Naturali, Rendicant, v. 40, no. 6, p. 1058–1065.

Perhac, R. M., 1964, Notes on the mineral deposits of the Gallinas Mountains, New Mexico, *in* Guidebook of the Ruidoso country: New Mexico Geological Society, 15th Field Conference, p. 152–154.

Pesty, L., 1966, Termeszetes szilikagel a Matra-hegysegbol: Foeldtani Koezloeny, v. 96, no. 2, p. 234–236.

Peterson, D. L., and Rambo, W. L., 1967, Bouguer gravity anomaly map of the Bearpaw Mountains and vicinity, Montana: U.S. Geological Survey Open-File Report (no longer available; see data in following reference).

—— , 1972, Principal facts for gravity stations in the Bearpaw Mountains and vicinity, Montana: U.S. Department of Commerce (Springfield, Virginia 22151), National Technical Information, serial PB2-10682, 15 p.

Petersen, H. W., 1979, Structural and petrological relationshps of the Bear Mountains intrusive, Silver City Range, Grant County, New Mexico [M.S. thesis]: Houston, University of Houston, 112 p.

Peyronnet, R., 1964, Etude de la bordure cristalline de la Limagne entre les vallees de la Tiretaine et de la Couze Chambon: Revue des Sciences Naturelles d'Auvergne, v. 30, part 1–4, p. 17–39.

Piccoli, G., and Spadea, P., 1964, Ricerche geologiche e petrografiche sul vulcanismo della Tripolitania settentrionale: Padua University, Instituto di Geologia e Mineralium Memoir 24, 68 p.

Pirsson, L. V., 1905, Petrography and geology of the igneous rocks of the Highwood Mountains, Montana: U.S. Geological Survey Bulletin 237, 208 p.

Plouff, D., 1958, Regional gravity survey of the Carrizo Mountains area, Arizona and New Mexico: U.S. Geological Survey Open-File Report, 30 p.

Plouff, D., and Pakiser, L. C., 1972, Gravity study of the San Juan Mountains, Colorado: U.S. Geological Survey Professional Paper 800B, p. B183–B190.

Poka, T., and Simo, B., 1966, A mellekkozet szerepe a Nagybatony kornyeki szubuulkani kepzodmenyek kialakulasaban: Foldtani Kozlony, v. 96, no. 4, p. 441–452.

Pollard, D. D., 1969, Deformation of host rocks during sill and laccolith formation [Ph.D. thesis]: Stanford, Stanford University, 134 p.

—— , 1972, Elastic-plastic bending of strata over a laccolith; Why some laccoliths have flat tops [abs.]: EOS, Transactions, American Geophysical Union, v. 53, no. 11, p. 1117.

—— , 1973, Derivation and evaluation of a mechanical model for sheet intrusions: Tectonophysics, v. 19, p. 233–269.

—— , 1976, On the form and stability of open hydraulic fractures in the earth's crust: Geophysical Research Letters, v. 3, no. 9, p. 513–516.

Pollard, D. D., and Holzhausen, G., 1979, On the mechanical interaction between a fluid-filled fracture and the earth's surface: Tectonophysics, v. 53, no. 1–2, p. 27–57.

Pollard, D. D., and Johnson, A. M., 1969, Sill-laccolith-bysmalith; Evolution of concordant intrusions in the Henry Mountains of Utah: Geological Society of America Abstracts with Programs, v. 2, p. 180.

—— , 1973, Mechanics of growth of some laccolithic intrusions in the Henry Mountains, Utah, Part II; Bending and failure of overburden and sill formation: Tectonophysics, v. 18, no. 3–4, p. 311–354.

Pollard, D. D., and Muller, O. H., 1976, The effect of gradients in regional stress and magma pressure on the form of sheet intrusions in cross section: Journal of Geophysical Research, v. 81, no. 5, p. 975–984.

Pollard, D. D., Muller, O. H., and Dockstader, D. R., 1975, The form and growth of fingered sheet intrusions: Geological Society of America Bulletin, v. 86, no. 3, p. 351–363.

Powers, S., 1916, Intrusive bodies at Kilauea: Zeitschrift fur Vulkanologie, v. 3, p. 28–35.

—— , 1921, Solitario uplift, Presidio-Brewster Counties, Texas: Geological Society of America Bulletin, v. 32, no. 4, p. 417–428.

Prager, W., and Hodge, P. G., Jr., 1951, Theory of perfectly plastic solids: New York, Dover, 264 p.

Prandtl, L., 1920, Ueber die Haerte plasticher Koerper: Akademie der Wissenschaften in Gottingen, Mathematisch-Physikalische Klasse, Nachrichten, K1, p. 74–85.

Pratt, W. P., and Jones, W. R., 1961, Trap-door intrusion of the Cameron Creek laccolith, near Silver City, New Mexico: U.S. Geological Survey Professional Paper 424-C, p. C164–C167.

———, 1965, The Cameron Creek laccolith; A trap-door intrusion near Silver City, New Mexico, *in* Guidebook of southwestern New Mexico, Part II: New Mexico Geological Society, 16th Field Conference, p. 158–163.

Raina, V. K., 1974, Zoned basic-ultrabasics of South Andaman Island: Indian Minerals, v. 28, no. 3, p. 99–101.

Ramberg, H., 1970, Model studies in relation to intrusion of plutonic bodies, *in* Newall, G., and Rast, N., eds., Mechanism of igneous intrusion: Liverpool, Gallery Press, p. 261–286.

———, 1981, Gravity, deformation, and the Earth's crust (second edition): New York, Academic Press, 452 p.

Ransome, F. L., 1909, Geology and ore deposits of Goldfield, Nevada: U.S. Geological Survey Professional Paper 66, 155 p.

Rao, S. S., 1967, The geological investigations of the Girnar igneous complex, Gujarat state: Mineral Wealth (Gujarat, Directorate of Geology and Mining), v. 3, no. 4, p. 10–14.

Read, H. H., Phemister, J., and Ross, G., 1926, The geology of Strath Oykell and lower Lock Shin: Scotland Geological Survey Memoir, p. 83–86.

Reeves, F., 1924, Structure of the Bearpaw Mountains, Montana: American Journal of Science, 5th Series, v. 8, p. 296–311.

———, 1925, Geology and possible oil and gas resources of the faulted area south of the Bearpaw Mountains, Montana: U.S. Geological Survey Bulletin 751-C, p. 71–114.

———, 1929, Thrust faulting and oil possibilities in the plains adjacent to the Highwood Mountains, Montana: U.S. Geological Survey Bulletin 806-E, p. 155–190.

———, 1931, Geology of the Big Snowy Mountains, Montana: U.S. Geological Survey Professional Paper 165, p. 135–149.

Reid, A. B., 1980, Aeromagnetic survey design: Geophysics, v. 45, no. 5, p. 973–976.

Reynolds, D. L., 1937, The Shonkin Sag laccolith (a discussion): American Journal of Science, 5th Series, v. 33, no. 202, p. 314–315.

Rich, F. J., ed., 1985, Geology of the Black Hills, South Dakota and Wyoming (second edition): American Geological Institute, 295 p.

Robinson, G. D., 1963, Geology of the Three Forks Quadrangle, Montana, *with a section on* Petrography of igneous rocks, by H. F. Barnett: U.S. Geological Survey Professional Paper 370, 143 p.

Robinson, H. H., 1913, The San Franciscan volcanic field, Arizona: U.S. Geological Survey Professional Paper 76, 213 p.

Robinson, C. S., Mapel, W. J., and Bergendahl, M. H., 1964, Stratigraphy and structure of the northern and western flanks of the Black Hills uplift, Wyoming, Montana, and South Dakota: U.S. Geological Survey Professional Paper 404, 134 p.

Rock, N.M.S., 1978, Petrology and petrogenesis of the Monchique alkaline complex, southern Portugal: Journal of Petrology, v. 19, part 2, p. 171–214.

Rockey, D. L., 1974, Geology of the eastern laccolith area, Meade County, South Dakota [M.S. thesis]: Rapid City, South Dakota School of Mines and Technology, 44 p.

Roemmel, J. S., 1982, A petrographic and economic evaluation of the White Cow intrusion, Little Rocky Mountains, Montana [M.S. thesis]: Missoula, University of Montana, 51 p.

Rose, A. W., 1966, Geology of part of the Amphitheater Mountains, Mt. Hayes Quadrangle, Alaska: Alaska Division of Mines and Minerals Geology Report no. 19, 12 p.

Ross, C. P., 1937, A sphenolith in the Terlingua district, Texas: Transactions, American Geophysical Union, 18th Annual Meeting, part 1, p. 255–258.

Rouse, J. T., 1933, The structure, inclusions, and alteration of the Deer Creek intrusive, Wyoming: American Journal of Science, 5th Series, v. 26, no. 152, p. 139–146.

Rouse, J. T., Hess, H. H., Foote, F., Vhay, J. S., and Wilson, K. P., 1937, Petrology, structure, and relations to tectonics of porphyry intrusions in the Beartooth Mountains, Montana: Journal of Geology, v. 45, no. 7, p. 717–740.

Runner, J. J., 1943, Structure and origin of the Black Hills pre-Cambrian granite domes: Journal of Geology, v. 51, no. 7, p. 431–457.

Ruppel, E. T., 1972, Geology of pre-Tertiary rocks in the northern part of Yellowstone National Park, Wyoming, *with a section on* Tertiary laccoliths in and near the Gallatin Range: U.S. Geological Survey Professional Paper 729-A, 66 p.

Rusu, D., 1975, The volcanic structures in gravimetric anomalies of the central and southern zones of Harghita Mountains: Volcanisme Bulletin, v. 38, no. 4, p. 1192–1204.

Sanford, A. R., and 7 others, 1977, Geophysical evidence for a magma body in the crust in the vicinity of Socorro, New Mexico, *in* Heacock, J. G., Keller, G. V., Oliver, J. E., and Simmons, G., eds., The earth's crust; Its nature and physical properties: American Geophysical Union Monograph 20, p. 385–403.

Santallier, D., and Floc'h, J. P., 1979, Magmatisme, geochronologie et tectonique dans le Massif Central: France, Bureau de Recherches Geologiques et Minieres, Bulletin, Serie 2, Section 1, no. 2, p. 109–119.

Savage, W. Z., 1974, Stress and displacement fields in stably folded rock layers [Ph.D. thesis]: College Station, Texas A&M University, 193 p.

Savage, W. Z., and Sowers, G. M., 1972, Criteria for separating stable and unstable folding III—Stable bending [abs.]: EOS, Transactions, American Geophysical Union, v. 53, p. 523.

Schenk, V., and Lattard, D., 1981, Formation and destruction of orthopyroxenes in granulites of the southern Calabrian Massif, Italy: Fortschritte der Mineralogie, Beiheft, v. 59, no. 1, p. 172–173.

Schmidt, R. G., Pecora, W. T., Bryant, B., and Ernst, W. G., 1961, Geology of the Loyd Quadrangle, Bearpaw Mountains, Blaine County, Montana: U.S. Geological Survey Bulletin 1081-E, p. 172–188.

Schmidt, R. G., Pecora, W. T., and Hearn, B., 1964, Geology of the Cleveland Quadrangle, Bearpaw Mountains, Blaine County, Montana: U.S. Geological Survey Bulletin 1141-P, p. P1–P26.

Schwan, W., 1968, Kinematics of certain major structures in the Variscan of northeastern Bavaria; Transverse dislocations in the Franconian forest, underthrusts in Thuringia, and the Munchberg: Oberrheinischer Geologischer Verein, Jahresberichte und Mitteilungen, v. 50, p. 127–141.

———, 1976, The intermediate massifs in the Saxothuringikum (German Variscan Orogene): Tectonophysics, v. 34, no. 1–2, p. 149–161.

Schwartz, G. M., and Davidson, D. M., 1952, Geologic setting of the copper-nickel prospect in the Duluth gabbro near Ely, Minnesota: Mining Engineering, v. 4, no. 7, p. 699–702.

Sclar, C. B., 1958, The Preston gabbro and the associated metamorphic gneisses, New London County, Connecticut: Connecticut Geological Natural History Survey Bulletin, no. 88, 136 p.

Scott, D. R., and Stevenson, D. J., 1986, Magma ascent by porous flow: Journal of Geophysical Research, v. 91, no. B9, p. 9283–9296.

Secor, D. T., and Pollard, D. D., 1975, On the stability of open hydraulic fractures in the earth's crust: Geophysical Research Letters, v. 2, p. 510–513.

Segalovich, V. I., 1971, Structure of the Kempirary ultrabasic massif: Akademiya Nauk S.S.R., Doklady, v. 198, no. 1, p. 178–181.

Shand, S. J., 1910, On borolanite and its associates in Assynt (2nd communication): Edinburgh Geological Society Transactions, v. 9, p. 376–419.

Shannon, W. M., and Goodell, P. C., 1983, A geochemical study of extreme differentiation; The Quitman caldera complex, West Texas: Geological Society of America Abstracts with Programs, v. 15, p. 391.

Shaw, H. R., 1980, The fracture mechanisms of magma transport from the mantle to the surface, *in* Hargraves, R. B., ed., Physics of magmatic processes: Princeton, University Press, p. 201–264.

Shaw, H. R., Wright, T. L., Peck, D. L., and Okamura, R., 1968, The viscosity of basaltic magma—An analysis of field measurements in Makaopuhi Lava Lake, Hawaii: American Journal of Science, v. 266, no. 4, p. 225–264.

Shepard, T. M., 1982a, Structural evolution and economic potential of the Rosillos Mountain area, Trans-Pecos Texas: Geological Society of America Abstracts with Programs, v. 14, no. 3, p. 136.

—— , 1982b, Geology of the Rosillos Mountains, Trans-Pecos Texas [M.A. thesis]: Fort Worth, Texas Christian University, 135 p.

Shoemaker, E. M., and Newman, W. L., 1953, Ute Mountains, a laccolithic feature in southwestern Colorado: Geological Society of America Bulletin, Abstracts, v. 64, no. 12, part 2, p. 1555.

Simmons, M. G., 1984, Contact metamorphism at the Cameron Creek laccolith, Grant County, New Mexico [M.S. thesis]: Monroe, Northeast Louisiana University, 58 p.

Simpson, D. G., 1980, A centrifuge model study of the evolution of laccolithic intrusions [M.S. thesis]: Kingston, Ontario, Queens University, 199 p.

Singewald, Q. D., 1942, Stratigraphy, structure, and mineralization in the Beaver-Tarryall area, Park County, Colorado: U.S. Geological Survey Bulletin 928-A, p. 1–44.

Sneddon, I. N., 1946, The distribution of stress in the neighborhood of a crack in an elastic solid: Royal Society of London, Proceedings, Series A, v. 187, p. 229–260.

Sneddon, I. N., and Lowengrub, M., 1969, Crack problems in the classical theory of elasticity: New York, John Wiley and Sons, 221 p.

Sofranoff, S. E., 1979, Geology, alteration, and mineralization of the carbonate mining district and surrounding area, Lawrence County, South Dakota [M.S. thesis]: Rapid City, South Dakota School of Mines and Technology, 150 p.

—— , 1981, Tertiary igneous development of the carbonate Spearfish Peak area, Lawrence County, South Dakota: Geological Society of America Abstracts with Programs, v. 13, no. 4, p. 226–227.

Sotteck, T. C., 1959, Geology of the Deadman Mountain and Whitewood anticline area, Mead-Lawrence Counties, South Dakota [M.S. thesis]: Rapid City, South Dakota School of Mines and Technology, 41 p.

Spear, D. B., 1977, Big Southern Butte, a silicic dome complex on the eastern Snake River plain, Idaho: Geological Society of America Abstracts with Programs, v. 9, no. 6, p. 765–766.

Spear, D. B., and King, J. S., 1982, The geology of Big Southern Butte, Idaho, *in* Bonnichsen, B., and Breckenridge, R. M., eds., Cenozoic geology of Idaho: Idaho Bureau of Mines and Geology Bulletin 26, p. 395–403.

Speed, R. C., 1976, Geologic map of the Humboldt lopolith and surrounding terrane, Nevada: Geological Society of America Map and Chart Series MC-14, scale 1:80,000, 4 p.

Spence, D. A., and Turcotte, D. L., 1985, Magma driven propagation of cracks: Journal of Geophysical Research, v. 90, no. B1, p. 575–580.

Spera, F. J., Yuen, D. A., Greer, J., and Sewell, G., 1985, Magma withdrawal from compositionally zoned magma chambers [abs.]: EOS, Transactions, American Geophysical Union, v. 66, no. 18, p. 397.

Spurney, J. C., 1982, The origin and characteristics of magnetite in the Iron Peak laccolith, Markagunt Plateau Utah: Geological Society of America Abstracts with Programs, v. 14, no. 6, p. 350.

Spurr, J. E., Garrey, G. H., and Fenner, C. N., 1912, Study of a contact metamorphic ore deposit: Economic Geology, v. 7, p. 444–484.

Squarci, P., and Taffi, L., 1963, Geologia della zona di Chiani - Laiatico - Orciatico (provincia di Pisa): Italian Geological Society Bulletin, v. 82, no. 2, p. 219–290.

Stark, M., 1907, Formen und genese lakkolithischer intrusionen: Festschrift der Naturwissenschaftliche Verein der Akademie Wien, p. 52–66.

—— , 1912, Beitrage zum geologisch-petrographischen Aufban der Euganeen und zur lakkolithen frage: Tschermaks Mineralogische und Petrographische Mitteilungen, neue folge, p. 80.

Steinmann, G., 1910, Gebirgsbildung und Massengestine in der Kordillere Südamerickas: Geologische Rundschau, Band 1, p. 13–35.

Stenzel, H. B., 1936, Structural study of a phacolith: International Geological Congress, 16th session, Washington, 1933 Report, v. 1, p. 361–367.

Stephansson, O., and Berner, H., 1971, The finite element method in tectonic processes: Physics of the Earth and Planetary Interiors, v. 4, p. 301–321.

Stephens, T., 1902, Notes on the diabase of Tasmania and its relation to the sedimentary rocks with which it is associated: Australian Association to Advance Science, v. 9, p. 251–263.

Stephenson, E. L., 1940, The results of magnetometer surveys on laccoliths in the Highwood Mountains, Montana [abs.]: Washington Academy of Science Journal, v. 30, no. 10, p. 488–489.

Stephenson, R., and Thomas, M. D., 1979, Three-dimensional gravity analysis of the Kiglapait layered intrusion, Labrador: Canadian Journal of Earth Sciences, v. 16, no. 1, p. 24–37.

Stewart, D. B., Pecora, W. T., Engstrom, D. B., and Dixon, H. R., 1957, Preliminary geologic map of the Centennial Mountain Quadrangle, Bearpaw Mountains, Montana: U.S. Geological Survey Miscellaneous Geological Investigations Map I-235, scale 1:31,680.

Storms, W. H., 1899, Laccoliths and their relations to ore deposits: Mining Science Press, v. 79, p. 745 (continued in v. 80, p. 5–6).

Strobell, J. D., Jr., 1956a, Geology of the Carrizo Mountains area, Arizona–New Mexico [Ph.D. thesis]: New Haven, Yale University, 45 p.

—— , 1956b, Geology of the Carrizo Mountains area in northeastern Arizona and northwestern New Mexico: U.S. Geological Survey Oil and Gas Investigations Map OM-160, scale 1:48,000.

Sullivan, K. R., 1987, Isotopic ages of igneous intrusions in southeastern Utah; Evidence for a mid-Cenozoic Reno - San Juan magmatic zone [M.S. thesis]: Provo, Brigham Young University, 15 p.

Summerhayes, C. P., 1968, Bathymetry and topographic lineation in the Cook Islands: New Zealand Journal of Geology and Geophysics, v. 10, no. 6, p. 1382–1399.

Sviridenko, V. T., 1976, The rapakivi granite formation of the western part of the Aldan Shield: International Geology Review, v. 18, no. 9, p. 1084–1096.

Swick, C. H., 1942, Pendulum gravity measurements and isostatic reductions: National Geodetic Survey, Special Publication no. 232, 82 p.

Tabor, R. W., and Crowder, D. F., 1969, On batholiths and volcanoes—intrusion and eruption of late Cenozoic magmas in the Glacier Peak area, North Cascades, Washington: U.S. Geological Survey Professional Paper 604, 67 p.

Takahashi, K., 1970, Geochemical investigations of the titaniferous iron-ore deposits, Wadi Hayyan and Wadi Qabqab area: Saudi Arabia, Director General Mineral Resources, Mineral Resources Research 1969, p. 74–79.

Taubeneck, W. H., and Poldervaart, A., 1960, Geology of the Elkhorn Mountains, northeastern Oregon, Part 2, Willow Lake intrusion: Geological Society of America Bulletin, v. 71, p. 1295–1322.

Tedlie, W. D., 1960, Acid rocks associated with an intrusive complex, Coppermine River area, Northwest Territories [M.S. thesis]: Vancouver, University of British Columbia, 57 p.

Thompson, T. B., 1964, The geology of the South Mountain area, Bernalillo, Sandoval, and Santa Fe Counties, New Mexico [abs.], *in* Guidebook of the Ruidoso country: New Mexico Geological Society, 15th Field Conference, New Mexico Bureau of Mines and Mineral Resources, p. 188.

Thorpe, M. R., 1919, Structural features of the Abajo Mountains, Utah: American Journal of Science, 4th Series, v. 48, p. 379–389 (for petrography see p. 80–84).

—— , 1938, Structure of the Abajo Mountains, *in* Gregory, H. E., ed., The San Juan country: U.S. Geological Survey Professional Paper 188, p. 89–91.

Trechmann, C. T., 1942, Metasomatism and intrusion in Jamaica: Geological Magazine, v. 79, no. 3, p. 161–178.

Turcotte, D. L., 1982, Magma migration: Annual Reviews of Earth and Planetary Science, v. 10, p. 397–408.

Turcotte, D. L., and Emerman, S. H., 1985, Magma fracture as a mechanism for magma migration [abs.]: EOS, Transactions, American Geophysical Union, v. 66, no. 18, p. 361.

Twenhofel, W. H., 1926, Intrusive granite of the Rose dome, Woodson County, Kansas: Geological Society of America Bulletin, v. 37, p. 402–412.

Twenhofel, W. H., and Bremer, B., 1928, An extension of the Rose dome intrusive, Kansas: American Association of Petroleum Geology, v. 12, p. 757–762.

Tyrrell, G. W., 1909, Geology and petrology of the intrusions of the Kilsyth-Croy district, Dumbartonshire: Geological Magazine, Decade 5, v. 6, p. 359–366.

——, 1928, The geology of Arran: Geological Survey of Scotland Memoir, 292 p.

Usiriprisan, C., 1979, Geology of the Woodville Hills intrusive, Lawrence County, South Dakota [M.S. thesis]: Rapid City, South Dakota School of Mines and Technology, 74 p.

Ussing, N. V., 1911, Geology of the country around Julianehaab, Greenland: Copenhagen, Meddelelser om Grønland, v. 38, 368 p.

Van Hise, C. R., and Leith, C. K., 1911, The geology of the Lake Superior region: U.S. Geological Survey Monograph 52, 641 p.

Vashchilov, Yu. Ya., 1963, Glubinnyye razlomy yuga Yano-Kolymskoy skladchatoy zony i ikh rol' v obrazovanii granitnykh intruziv i formirovanii struktur: Soviet Geology, no. 4, p. 54–72.

Vauchelle, L., and Lemeyre, J., 1983, The western part of the Gueret Massif (French Central Massif); Lithology, overall structure, and mineralization: Comptes Rendu de l'Academie des Sciences, Series 2, Mecanique-Physique, Chimie, Sciences de l'Univers, Sciences de la Terre, v. 297, no. 1, p. 63–68.

Verbeek, K., 1971, The mechanism of emplacement of the Marble Mountain laccolith [M.S. thesis]: State College, Pennsylvania State University, 112 p.

——, 1972, The mechanism of emplacement of the Marble Mountain laccolith, Flagstaff, Arizona: Plateau, v. 45, no. 2, p. 68–72.

Viola, C., 1892, Nota preliminaire sulla regione dei gabbri e delle serpentine nell' alta valle del Sinni in Basilicata: Bollettino del Roma Comitato geologico d'Italia, v. 23, p. 105.

Waagé, K. M., 1959, Stratigraphy of the Inyan Kara Group in the Black Hills: U.S. Geological Survey Bulletin 1081-B, p. 11–90.

Waff, H. S., 1986, Introduction to special section on partial melting phenomena in earth and planetary evolution: Journal of Geophysical Research, v. 91, no. B9, p. 9217–9221.

Wager, L. R., and Brown, G. M., 1967, Layered igneous rocks: San Francisco, W. H. Freeman and Company, 588 p.

Wagle, B. G., and Almeida, F., 1974, Mega-porphyritic dolerite intrusion at San Pedro, Goa: Current Science, v. 43, no. 12, p. 378–380.

Washington, H. S., 1901, The Magnet Cove laccolith, Arkansas [abs.]: New York Academy of Sciences, Annal 13, p. 448–449.

Watson, E. H., 1937, The geology and biology of the San Carlos Mountains, Tamaulipas, Mexico, Part 2, Igneous rocks of the San Carlos Mountains: Michigan University Studies of Science Series, v. 12, p. 99–156.

Watts, W. W., 1886, The Corndon laccolites: British Association for Advancement of Science, Report, p. 670.

Weed, W. H., 1899a, Fort Benton, Montana: U.S. Geological Survey Geological Atlas Folio GF-55, 9 p.

——, 1899b, Laccoliths and bysmaliths [abs.]: Science, new series, v. 10, p. 25–26.

——, 1900, Geology of the Little Belt Mountains, Montana: U.S. Geological Survey, 20th Annual Report, part 3, p. 257–461.

——, 1901, Geology and ore deposits of the Elkhorn mining district, Jefferson County, Montana: U.S. Geological Survey, 22nd Annual Report, part 2, p. 399–550.

Weed, W. H., and Pirsson, L. V., 1895a, Highwood Mountains of Montana: Geological Society of America Bulletin, v. 6, p. 389–422.

——, 1895b, On the igneous rocks of the Sweet Grass Hills, Montana: American Journal of Science, 3rd Series, v. 50, p. 309–313.

——, 1896a, Igneous rocks of the Bearpaw Mountains, Montana: American Journal of Science, 4th Series, v. 1, p. 283–301, 351–362; v. 2, p. 136–148, 188–199.

——, 1896b, Geology of the Little Rocky Mountains: Journal of Geology, v. 4, p. 399–428.

——, 1896c, Geology of the Castle Mountains mining district, Montana: U.S. Geological Survey Bulletin 139, 164 p.

——, 1898, Geology and mineral resources of the Judith Mountains of Montana: U.S. Geological Survey, 18th Annual Report, part 3, p. 437–616.

——, 1901, Geology of the Shonkin Sag and Palisade Butte laccoliths in the Highwood Mountains of Montana: American Journal of Science, 4th Series, v. 12, p. 1–17.

Weertman, J., 1971a, Theory of water-filled crevasses in glaciers applied to vertical magma transport beneath oceanic ridges: Journal of Geophysical Research, v. 76, no. 5, p. 1171–1183.

——, 1971b, Velocity at which liquid-filled cracks move in the earth's crust or in glaciers: Journal of Geophysical Research, v. 76, no. 35, p. 8544–8553.

——, 1973, Oceanic ridges, magma filled cracks and mantle plumes: Geofisica Internacional, v. 13, p. 317–336.

——, 1980, The stopping of a rising, liquid-filled crack in the earth's crust by a freely slipping horizontal joint: Journal of Geophysical Research, v. 85, no. B2, p. 967–976.

Weertman, J., and Weertman, J. R., 1964, Elementary dislocation theory: New York, Macmillan, 213 p.

White, S. F., 1980, Petrology of the Cenozoic igneous rocks of the Lytle Creek area, Bear Lodge Mountains, Wyoming [M.S. thesis]: Grand Forks, University of North Dakota, 69 p.

Whiting, C. K., 1977a, Trachydolerite laccolith and feeder dikes, northern Adel Mountains, Cascade County, Montana: Geological Society of America Abstracts with Programs, v. 9, no. 6, p. 773–774.

——, 1977b, Small laccoliths and feeder dikes of the northern Adel Mountain volcanics [M.S. thesis]: Missoula, University of Montana, 74 p.

Wiley, M. A., 1972, Gravity, magnetic, and generalized geologic map of the Van Horn–Sierra Blanca region, Trans-Pecos Texas: Austin, University of Texas, Bureau of Economic Geology, Geologic Quadrangle Maps with Text no. 40, 26 p.

Williams, H., 1929, Geology of the Marysville Buttes, California: Berkeley, California University, Department of Geological Sciences Bulletin, v. 18, no. 5, p. 103–220.

Williams, H., and McBirney, A. R., 1979, Volcanology: San Francisco, Freeman Cooper and Company, 397 p.

Wilshire, H. G., 1967, The Prospect alkaline diabase-picrite intrusion, New South Wales, Australia: Journal of Petrology, v. 8, p. 97–163.

Wise, H. M., 1977, Geology of the igneous intrusions of the northern Hueco Mountains: Geological Society of America Abstracts with Programs, v. 9, no. 1, p. 81.

Wisser, E., 1960, Relation of ore deposition to doming in the North American Cordillera: Geological Society of America Memoir 77, 117 p.

Witkind, I. J., 1957, Abajo Mountains, Utah: U.S. Geological Survey Report TEI-690 (book 1), p. 143–147.

——, 1958, The Abajo Mountains, San Juan County, Utah: Intermountain Association of Petroleum Geologists, 9th Annual Field Conference Guidebook, p. 60–65.

——, 1964a, Geology of the Abajo Mountains area, San Juan County, Utah: U.S. Geological Survey Professional Paper 453, 110 p.

——, 1964b, Preliminary geologic map of the Tepee Creek Quadrangle, Montana–Wyoming: U.S. Geological Survey Miscellaneous Geological Investigations Map I-417, scale 1:48,000.

——, 1964c, Age of the grabens in southeastern Utah: Geological Society of America Bulletin, v. 75, no. 2, p. 99–105.

——, 1965a, Fracture-controlled laccoliths, Abajo Mountains, Utah [abs.]: Geological Society of America Special Paper 82, p. 351–352.

——, 1965b, Relation of laccolithic intrusion to faulting in the northern part of the Barker Quadrangle, Little Belt Mountains, Montana, *in* Geological Survey Research 1965: U.S. Geological Survey Professional Paper 525-C, p. C20–C24.

——, 1966, Structural framework of the north part of the Barker Quadrangle, Little Belt Mountains, Montana [abs.]: Geological Society of America Special Paper 87, p. 306.

——, 1969, Geology of the Tepee Creek Quadrangle, Montana–Wyoming: U.S. Geological Survey Professional Paper 609, 101 p.

——, 1971, Geologic map of the Barker Quadrangle, Judith Basin and Cascade Counties, Montana: U.S. Geological Survey Geological Quadrangle Maps GQ-898, scale 1:62,500.

—— , 1973, Igneous rocks and related mineral deposits of the Barker Quadrangle, Little Belt Mountains, Montana: U.S. Geological Survey Professional Paper 752, 58 p.

—— , 1975, The Abajo Mountains; An example of the laccolithic groups on the Colorado Plateau, Canyonlands County: Four Corners Geological Society, Field Conference Guidebook 8, p. 245–252.

Wolff, J. E., 1938, Igneous rocks of the Crazy Mountains, Montana: Geological Society of America Bulletin, v. 49, no. 10, p. 1569–1626.

Woolley, A. R., 1970, The structure relationships of the Loch Borrolan complex, Scotland: Geological Journal, v. 7, part 1, p. 171–182.

Wynn, J. C., and Bhattacharya, B. K., 1977, Reduction of terrain-induced aeromagnetic anomalies by parallel-surface continuation; A case history from the southern San Juan Mountains, Colorado: Geophysics, v. 42, no. 7, p. 1431–1449.

Yagi, K., 1953, Recent activity of Usu volcano, Japan, with special reference to the formation of Syowa Sinzan: Transactions, American Geophysical Union, v. 34, no. 3, p. 449–456.

Yang, H. Y., and Lee, R. F., 1978, Mineralogy and petrology of Fuhsing hypabyssal suite: Proceedings of the Geological Society of China, v. 21, p. 43–66.

Yates, R. G., and Thompson, G. A., 1959, Geology and quicksilver deposits of the Terlingua district, Texas: U.S. Geological Survey Professional Paper 312, 114 p.

Young, W. E., 1947, Iron deposits, Iron County, Utah: U.S. Bureau of Mines Report of Investigations 4076, 102 p.

Zhidkov, A. Y., 1978, Classification and distribution of potassium-aluminum silicate ores in the Synnyr alkalic pluton: Doklady of the Academy of Sciences of the U.S.S.R., Earth Science Sections, v. 242, no. 1-6, p. 107–110.

Zienkiewicz, O. C., 1977, The finite element method (third edition): New York, McGraw-Hill, 787 p.

MANUSCRIPT ACCEPTED BY THE SOCIETY SEPTEMBER 22, 1987

Typeset by WESType Publishing Services, Inc., Boulder, Colorado
Printed in U.S.A. by Malloy Lithographing, Inc., Ann Arbor, Michigan